PENGUIN CANADA

WAITING FOR THE MACAWS

TERRY GLAVIN is a renowned writer and conservationist. His book *The Last Great Sea: A Voyage Through the Human and Natural History of the North Pacific Ocean* won the Hubert Evans Prize, and *This Ragged Place: Travels Across the Landscape* was a Governor General's Award finalist. A frequent contributor to newspapers and magazines, Glavin is the recipient of numerous regional and national journalism awards. He is currently an adjunct professor at the University of British Columbia's fine arts department. He lives in Victoria, British Columbia.

Also by Terry Glavin

The Last Great Sea: A Voyage Through the Human and Natural History of the North Pacific Ocean

This Ragged Place: Travels Across the Landscape

Dead Reckoning: Confronting the Crisis in Pacific Fisheries

A Ghost in the Water

A Death Feast in Dimlahamid

Praise for *Waiting for the Macaws*

"*Waiting for the Macaws* asks us to care, deeply, about living in the midst of the greatest extinction rates of the past 65 million years. If there's room for hope, it can be found in a book like this, a book suffused with an urgency that is never haranguing, with an approach immersed in the human need to connect and transform through story."

—*The Globe and Mail*

"Terry Glavin is a wise and eloquent writer whose clear-eyed intelligence explores our conflicted relationship with nature and our fellow man. In *Waiting for the Macaws* he shows how we have shaped and disrupted the world we depend on. What Glavin has to tell is urgent, important, and well said."

—Ronald Wright, author of *A Short History of Progress*

"Glavin is an amiable companion on this around-the-world expedition … The reader is fortunate to be joined on the voyage by so lyrical an essayist."

—*Canadian Geographic*

"Glavin is one of the prophets of our time. He is able to see things that others do not or will not see, and then put together these disparate pieces to make a new whole. Not only can he see them but he can spin them into stories that speak to the deepest, most primal parts of the human brain."

—*Literary Review of Canada*

"*Waiting for the Macaws* is a haunting reminder of the scale and breadth of what can only be described as a catastrophe of the human spirit and imagination. In almost every case of extinction cited in this book, people play the pivotal and destructive role. This is both a discouraging and a hopeful observation, for it suggests that if human beings are the agents of destruction, we can also most surely be the facilitators of cultural and biological survival. Terry Glavin's remarkable book leaves little doubt that this is indeed the central challenge of our times."

—Wade Davis, author of *Light at the Edge of the World: A Journey Through the Realm of Vanishing Cultures*
and Explorer-in-Residence, National Geographic Society

"Glavin is a thoughtful and profoundly articulate conservationist whose passion has taken him around the globe. *Macaws* exhibits not just the commitment we'd expect of an activist's agenda, but also the refined humanism of fine travel writing." —*Edmonton Journal*

"One of the best in-depth journalists working outside the mainstream. His work reveals a formidable intellectual ability to discern the big picture in the smallest events. And he's developed an eloquent literary style that enables him to present complicated issues in plain, unadorned language but to deliver them using powerful images of poetic intensity."

—*The Vancouver Sun*

"Elegantly written, with his conservationist's heart shining throughout."

—*Burnaby Now*

"*Waiting for the Macaws* is an urgent, necessary book about the hells of our own making and the scraps of heaven that survive. Terry Glavin writes with both passion and authority about the fading diversity on everything from whales and apples to languages and belief systems. Do yourself and this struggling world a favour: let this book kindle your intellect. Let it break your heart. Let it stir your soul."

—Mark Abley, author of *Spoken Here: Travels Among Threatened Languages*

"A powerful odyssey of environmentalism." —*Ottawa Citizen*

"*Waiting for the Macaws* is a thoughtful look at the state of the world and the countless ways humans change it. The situation is bad, but it's not yet lost if we start making more sustainable decisions in the next decade or so. The book deserves to be chewed as slowly as the whale blubber Glavin at one point describes eating." —*Monday Magazine* (Victoria)

"Glavin's immersion in the most important story of our time lets us not just know but also feel that the loss of life and of ways of life is beyond appalling; it is threatening. He takes us to the heart of the matter by writing with heart."

—Richard Manning, author of *Against the Grain: How Agriculture Has Hijacked Civilization*

Waiting for the Macaws

and other stories from the age of extinctions

Terry Glavin

PENGUIN
CANADA

PENGUIN CANADA

Published by the Penguin Group

Penguin Group (Canada), 90 Eglinton Avenue East, Suite 700, Toronto, Ontario, Canada M4P 2Y3
(a division of Pearson Canada Inc.)

Penguin Group (USA) Inc., 375 Hudson Street, New York, New York 10014, U.S.A.
Penguin Books Ltd, 80 Strand, London WC2R 0RL, England
Penguin Ireland, 25 St Stephen's Green, Dublin 2, Ireland (a division of Penguin Books Ltd)
Penguin Group (Australia), 250 Camberwell Road, Camberwell, Victoria 3124, Australia
(a division of Pearson Australia Group Pty Ltd)
Penguin Books India Pvt Ltd, 11 Community Centre, Panchsheel Park, New Delhi – 110 017, India
Penguin Group (NZ), cnr Airborne and Rosedale Roads, Albany, Auckland 1310, New Zealand
(a division of Pearson New Zealand Ltd)
Penguin Books (South Africa) (Pty) Ltd, 24 Sturdee Avenue, Rosebank, Johannesburg 2196, South Africa

Penguin Books Ltd, Registered Offices: 80 Strand, London WC2R 0RL, England

First published in a Viking Canada hardcover by Penguin Group (Canada),
a division of Pearson Canada Inc., 2006
Published in this edition, 2007

(WEB) 10 9 8 7 6 5 4 3 2 1

Manufactured in Canada.

ISBN-10: 0-14-301657-1
ISBN-13: 978-0-14-301657-1

Library and Archives Canada Cataloguing in Publication data available upon request

Visit the Penguin Group (Canada) website at **www.penguin.ca**

Special and corporate bulk purchase rates available; please see
www.penguin.ca/corporatesales or call 1-800-810-3104, ext. 477 or 474

This book is dedicated to Vicky Husband, conservation chair of the Sierra Club of British Columbia, one of Canada's greatest conservationists, and a dear friend. This book is also dedicated to the millions of people all over the world who continue in the struggle to strengthen the things that remain.

Everything one writes now is overshadowed by this ghastly feeling that we are rushing towards a precipice and, though we shan't actually prevent ourselves or anyone else from going over, must put up some sort of a fight.

—George Orwell,
in a letter to Cyril Connolly,
December 14, 1938

Contents

Acknowledgments

I owe a great debt to my family for their patience and humour—my wife, Yvette, and my children, Zoe, Eamonn, and Conall. I especially thank Robert Harlow and Ben Parfitt for helping me with some difficult parts of the manuscript. Penguin's Susan Folkins was a marvellous and patient editor, and my agent, Jackie Kaiser, stood with me through it all. I owe the British Columbia Arts Council my gratitude for its early support, and I particularly thank Louise Pedersen for going beyond the call of duty as a research assistant when we were getting this project off the ground. Small portions of this book have appeared in essay form in *The Vancouver Review, Canadian Geographic, The Georgia Straight,* and *Adbusters,* and I thank the editors of those magazines for allowing me to test some of the ideas that have ended up in these pages.

In Ireland, my cousin Christine Gurnett provided far more insight than she will ever know. My old friend Marc Edge made me feel at home in Singapore. Janice D. Boyd, president of Amigos de las Aves USA, pointed me toward the macaws at the Curú refuge in Costa Rica, and Greg Matuzak welcomed me there. Without Dave Martin and Xan Augerot at the Wild Salmon Center in Portland I would never have made it to Russia, and without Vera Kharberger's gifts as a translator I would have been hopeless there. In the Lofotens, the Frovik and Bendiksen families of Reine showed me great kindness, as did the Ramberg family at the Lofoten Sjøhus og Rorbuer in Sorvagen.

For "The Ghost of the Woods," David Parker of Port Alice was especially generous with his time, and his help is deeply appreciated, given the pain of reliving the cougar attack. The University of Arizona's Paul Beier, Nova Scotia wildlife officer Mark Pulsifer, David Shackleton of the University of British Columbia and his graduate students, and British Columbia wildlife biologist Doug Janz and conservation officer Gerry Brunham were particularly helpful in assembling background information for that chapter.

Mark Lauckner, Tina Farmilo, Jennifer Iredale, and Brian Brett were very gracious in helping to assemble research materials about food-crop diversity. The staff at the Economic Botany collections at the Royal Botanical Gardens at Kew were especially helpful.

In Nagaland, I am forever indebted to the cleric and theologian Reverend C. Walu Walling of Impur; General "Jimmy" Singh and the staff at Gurudongma; Neisatuo Keditsu, president of the Nagaland Tourism Association; Kevi Meyase of the Army of Khonoma; my comrades, translators, and guides Khrienuo Kense, Bilong Nagi, and Visevor Nagi, as well as Ao Naga folklorist and singer Sangyusang Pongener. Alec Coupe, a linguist at Australia's La Trobe University, took time out of his busy schedule to help me understand the hidden "song language" known to Ao speakers.

And lastly, my thoughts are always with Sibu Das, of Calcutta, and his family.

A village

Prologue
The Valley of the Black Pig

The townland of Coolreagh is a place of rolling hills, bogs, and woods within the parish of Bodyke, in the northeastern corner of County Clare, in the Republic of Ireland. Its name comes from Cúl Riabhach, which can be rendered in English along the lines of The Grey Corner. Thickly hedged, stone-lined boreens connect the old farmhouses to one another and to the nearby villages and the outside world. Down one of these little roads is my mother's family's farm. The farmhouse stands on the banks of a stream called the Annamullaghaun, which means Mill River. The name acknowledges the farmhouse, which was once a small mill. The house has stood there in one form or another for 400 years.

I had come for a visit with my Uncle Tony and my Aunt Angela, and my cousins Christine, Philip, Douglas, and the rest. I'd just begun working on this book. It was my intention to make sense of some of my notes and then start out on a long walk.

In one of my notebooks I'd jotted down a passage from the Old Testament's Book of Hosea: *Therefore shall the land mourn, and everyone that dwelleth therein shall languish with the beasts of the field and with the fowls of heaven; yea, the fishes of the sea also shall be taken away.* There was also an article out of *The Guardian* that I'd scribbled on. It was from the May 23, 2002, edition, under the headline "The Way We'll Live in 2032": "The destruction of 70 per cent of the natural world in 30 years, mass extinction of species, and the collapse of human society in many countries is forecast in a bleak report by 1,100 scientists published yesterday."

From the farmhouse at Coolreagh, you can walk over a small stone bridge that crosses the Annamullaghaun beside the farm, and you will find yourself standing in the townland of Caherhurley beside a field called the Castle Field, which takes its name from a craggy and vine-covered rock in the middle of it, the remnant of a stone fort built by local tribesmen loyal to Brian Boru, the great warrior-chief who defeated the Vikings at Clontarf in 1014. In the Castle Field you will notice the ground beginning to rise gently, and if you walk that way, up Blackguard's Hill, you'll find yourself heading through Ballyvaughan into the Slieve Bernagh mountains.

If you walk in the other direction, northward, you will eventually find yourself in the townland of Fossamore, and the ground begins to rise there, too, into the Slieve Aughty mountains. It's wilder up that way. Above Fossamore is Powlagower, the Goat's Hole, and Tabernagat, the Well of the Cats. There is the Sruthánalunacht, the Stream of New Milk, which once ran white with milk but long ago it turned to water, they say, when a woman washed her feet in it. There are people who live at Cloonusker who say that at the end of the world, the final battle of the last war will be fought up there, above Gortaderra, in a place called the Valley of the Black Pig, and on that last day of battle the Stream of New Milk will turn to blood.

The story the old people at Cloonusker tell is the same event foretold by Hosea, in his way, and it is also the future imagined by those 1100 scientists, in their way, in that article in *The Guardian.* William

Butler Yeats was haunted by these things, and just as the world was carrying the great weight of dread and foreboding in his apocalyptic poem, *The Valley of the Black Pig,* so it was when I began writing this book.

This is a book about extinctions. It was written at the harsh dawn of an epoch that is coming to be called the Sixth Great Extinction. It is a time without parallel in the 65 million years that have passed since the end of the Cretaceous period. The world is again weary of empires. The dews still fall slowly, and the dreams still gather, but no matter *the clash of fallen horsemen and the cries of unknown perishing armies* in Yeats's poem, we wonder, and we wait, and we go about our business, even as the sound of something terrible slowly approaches from across the hills.

Roughly 34,000 plants, or 12.5 percent of all the plants known to science, are threatened with extinction. One in eight bird species is threatened with extinction, along with one in four mammals, one in three of all known amphibians, four of every ten turtles and tortoises, and half of all the surveyed fish species in the world's oceans, lakes, and rivers. Perhaps a million of our fellow species are trailing wearily toward that final cliff edge. We lose a distinct species, of one sort or another, every ten minutes.

These tabulations constitute only the most crude sort of barometer of the great unravelling of the living world, and the ecologists who calibrate extinction rates readily admit to that fact. But besides that, the greatest bleeding away of diversity is occurring well below the level of what geneticists and taxonomists consider a "species." It is also happening outside that category entirely. It is happening down where the true measure of life's diversity is found. Extinction is taking away the subspecies, the local population, the particular, the neighbourhood, the singular, and the specific. And it is not confining itself to the "wild" things of the world.

It is also carrying away the tamed—the functions of artificial selection. Estimates of the number of the world's vegetable varieties lost during the twentieth century run as high as 30,000, with another one vanishing every six hours. Of the thousands of apple varieties in North American orchards at the beginning of the twentieth century, for instance, all but one-seventh are gone. Of 2683 pear varieties, only slightly more than one-tenth remain. By the 1970s, most of what remained of Europe's old vegetable varieties were threatened with extinction. Even livestock breeds are disappearing—Europe lost half its distinct breeds during the twentieth century. Of those that are left, 43 percent are threatened with extinction.

Humanity's diversity is similarly withering. Though the world population has surpassed six billion, it is as though some savage ethnic cleansing is underway. The world is losing an entire language every two weeks. Fully half of the world's 5000 languages are expected to be gone, with all their songs and sagas, by the middle of this century. We are losing religious and intellectual traditions, entire bodies of literature, taxonomies, pharmacopias, and all those ways of seeing, knowing, and being that have made humanity so resilient and successful a species for so long. This is not what we had come to expect from the promise of the Enlightenment. We are not gaining knowledge with every human generation—we are losing it.

A dark and gathering sameness is descending upon the world, and the language of environmentalism is wholly inadequate to the task of describing the thing. It can't even come close. It isn't that environmentalism exaggerates the phenomenon. It's just that it doesn't have the words for it. By "environmentalism," I mean that great movement of people and ideas that emerged in the early 1970s, largely as a function of Euro-American liberalism. By "the language of environmentalism," I mean not just a narrative template that is overly burdened with outmoded ideas and cultural biases, but language that draws arbitrary distinctions between "wilderness" and everything else and that places "nature" outside of culture.

The simple premise of this book is that all these extinctions are related. The intent of this book is to explore the relationships among and between all these extinctions, and one cannot even begin to do that by relying on the prism of environmentalism. As a separate category of thought, environmentalism is of little use in comprehending the scale of extinctions the world is suffering. Extinction is the fate befalling the things Hosea described—the wild and the tamed, the land and everyone that dwelled therein, and the beasts of the field as well as the fowls of heaven. All these extinctions are part of the same phenomenon and properly part of the same conversation. The forces at work in the world are "not well understood," to borrow the vernacular of scientific journals. But they are *cultural* forces.

To make sense of the world, people tell stories. This is a book of stories, not only because, as novelist Doris Lessing says, "our brains are patterned for storytelling." It is also because at a time when the world is filled with dread and foreboding, and when the great master narratives we've relied on to understand things are collapsing all around us, there should be some virtue in going for a walk through the hills and coming back at the end of the day with an account—a story—of what's out there.

So that's what I attempted, and after a fairly thorough reconnaissance of the extinctions at work in the world, I found absolutely no evidence that any of this is what humanity really wants. That is good news. I can also confidently report that the roads and boreens that wind their way through the East Clare hills do not lead inevitably northward beyond Fossamore into the Valley of the Black Pig. Hosea's prophecy did not come to pass, the Israelites did not perish, and that *Guardian* article about the destruction of 70 percent of the natural world and the collapse of human societies within 30 years did not describe one inescapable fate. It was about a United Nations Environment Programme (UNEP) *Global Outlook* report that actually found four roads through the hills.

Only two of those roads lead into the Valley of the Black Pig. One of them, which the report calls a "security first" approach, traverses a desperate countryside of inequality and conflict, protest and reaction. The powerful and the wealthy end up gating themselves into enclaves, leaving the masses of poor to survive as best they can in the collapsing environment outside the walls. The other road takes a more circuitous route, but it ends up in the same place. The UNEP scientists call this road a "markets first" approach, because it requires us to put our faith solely in market forces, further globalization, and greater trade liberalization. We count on corporate ingenuity to resolve social and "environmental" problems and leave it to ethical investors and consumer groups to keep global capitalism honest. The state ends up with no capacity to regulate human affairs or protect those values that humanity cherishes.

The other two roads lead to a different sort of countryside altogether. A "policy first" approach leaves public governments in charge of identifying social and environmental goals. We put our trust in coordinated responses to environmental disruption and poverty, factor environmental and social costs in our public policy, and allow for local and regional innovation. On the "sustainability first" road, a wholly new paradigm emerges. New institutions make room for radical changes in the way people interact with one another and with the living, breathing world. Corporations are held to account. Citizens and stakeholder groups participate directly in decision making. We all muddle through.

The closer you look behind all those cringe-making global headlines, what you see is that practical solutions have already been found for almost all those dreadful problems the headlines describe. In many ways, the world is becoming a better place. The world is taking on another 200,000 people every day, but humanity's total numbers are projected to level off at around 10 billion before the year 2100. The last, critical redoubts of 43.8 percent of the world's vascular plants and more than one-third of the world's animals, birds, reptiles, and amphibians take up only about 1.5 percent of the earth's surface. If we take "wilderness" to mean tracts of land of at least 10,000 square kilometres with, say, at least 70 percent of the native ecosystems intact, then almost half the planet is still like that.

But fortune-telling has always been a dodgy enterprise. I grew up in a world crippled with the fear of nuclear apocalypse. Then the Berlin Wall fell. Then, just as the great gulf between capitalism and communism was finally breached, two passenger planes were plunged into those two great towers in New York, and the world was divided against itself again. No one foresaw the epochal changes to practically every aspect of our lives that occurred, almost overnight, as a consequence of computers and the internet. The exponential growth in information-technology capacity is moving from a curving line on a graph to a line that goes straight up, and no one knows where it's going. By the beginning of this century, the People's Republic of China had emerged as a great state-capitalist powerhouse with an economy so dynamic that it threatened to eclipse the United States, and the Muslim population of Western Europe was roughly the same as the population of Syria. No one foretold these things.

Even the meaning of the word *extinct* is blurring. Nature has always existed as much within the human imagination as anywhere else, but rapid advances in the genetic sciences, transgenic manipulation, and biotechnology are changing everything. They are throwing open the final borders between wild and domesticated animals and plants, humans and animals, "wilderness" and zoos, and humans and machines. The emerging fields of robotics and nanotechnology will lead us all to hell or to heaven—nobody seems to know which.

There are even people up in County Leitrim who make a convincing case that the people at Cloonusker haven't read St. Columba's prophecies correctly at all, and the real Valley of the Black Pig isn't that place above Gortaderra. It's actually up their way, they say, by Ballinamore.

It's true that the current extinction crisis is distinguished from the spasms that ended the Ordovician, Devonian, Permian, Triassic, and Cretaceous periods in that it can be reliably attributed, in one way or another, to a single species: *Homo sapiens.* But it is not a simple story, with human beings as the cruel villain of the piece. In the case against humanity, this book is offered as evidence for the defence.

The many creatures that have vanished during the course of the human experience make up a staggering and fantastic bestiary. There was an elephant-sized ground sloth in Texas, a Sicilian elephant no bigger than a pony, a bear-sized Australian wombat, and a Floridian bird with the wingspan of a Cessna. It is a long, sad, and continuing story, but it is certainly not solely a "modern" phenomenon. Most of the extinctions caused by human beings in the modern period have happened by accident, on account of culprits such as rats, pigs, goats, and other "introduced" species. For every sad and well-documented story of a species dying out because of humans, there is almost always an overlooked story of ordinary people struggling against the forces of extinction.

The immediate causes of extinction are often quite straightforward. The epicentre of the extinction of the world's "wild" things is the humid tropics, especially those parts of it in the thrall of the "markets first" and "security first" approaches the UNEP scientists described. Tropical extinctions are most often the result of a simple equation: Chop down all the trees and the animals will die. The temperate world is not at all like the tropics, but there are certain global patterns at work, patterns that involve upheaval and dispossession. Extinctions tend to follow the collapse of order in human societies. Human-caused extinctions are often the result of old "feedback loops" breaking down, and old restraints giving way. Where you have rapid advances in technology, dramatic shifts in political power, and profound economic disruption, extinction tends to follow. There is also a surprisingly direct correlation between the removal of vegetative cover—even "domesticated" plant life—and the dying out of languages, cultures, and ways of life.

It is not true that people started taking these things seriously only in the late 1960s, when astronauts beamed back pictures of this lovely blue planet, or when U.S. senator Gaylord Nelson conceived the idea of Earth Day. People have been taking these things seriously for a long, long time. Deep within the human consciousness is an ancient and abiding desire to be in the presence of flourishing, abundant, and diverse forms of life. Like the desire for narrative, enchantment with the beauty, utility, and diversity of living things is an inescapable aspect of human nature.

For all these reasons, when I started out on that long walk from the farm at Coolreagh, I was not drawn helplessly along the roads that lead up beyond Fossamore. I walked in a different direction, and I ended up at the Temple of Kali, in Calcutta. But I'm getting ahead of myself.

I set out with my cousin Christine, turning east at the top of the road from the farm. We walked in the direction of Lough Derg, the great broadening of the Shannon River that so deeply cleaves the country, and we headed toward the village of Tuamgraney, the Tomb of Gráinne. Some people associate Tuamgraney with Grían, an old sun goddess. Another way of translating the name of the village is Altar of the Sun. The fields and the stones argue among themselves about which of their stories is right and best, but all the old stories appear to agree on one thing: Gráinne was a high-born woman who became inconsolable and drowned herself in a little lake up above Feakle after learning she had been conceived of a sunbeam and would never know the world of mortal men.

Being mere mortals, when we pull away at the vines of our limited understanding of the sum of all living things, we still find the moss-covered foundation of Aristotle's *scala naturae* and the fading inscriptions chiselled by the seventeenth-century Lutheran medical student Carl von Linné (also known as Carolus Linnaeus), who gave us our system of taxonomic nomenclature, which he called the *systema naturae.* Nowadays, the whole edifice is crumbling. Entire families of species emerge from the mud, as though summoned by Zeus, or disappear forever, as though extinguished by the impact of a Cretaceous asteroid. Sometimes they vanish out of the known world, and sometimes they vanish owing to the mere publication of papers in such scholarly periodicals as the *Journal of Heredity.*

But the world of mere mortals will never be made up of species that fit neatly into their own genus, family, order, class, phylum, kingdom, and domain. The mortal world is also made up of stories. That's the first thing you notice when you walk through the East Clare hills. A single narrative is not so easily imposed on the land. Each townland is its own piece of a quilt. Within each townland there are fields. Each field accounts for itself in its own way, with its own stories.

The little field beside the farmhouse at Coolreagh is Carrigrua, the Red Crag. Across the boreen from Carrigrua is the Big Hollow. Above the Big Hollow is Hogan's. Then there's the tillage field, the old milking shed, and beyond the tillage field is Flanagan's. Jack Brian's field is all covered in whitethorns and holly and blackthorns. It's also called the Fairy Field, because within it is a rath, which is a ring fort, one of those circles of stones where people used to see faint lights dancing on certain nights of the year. The holy well at Tobar Coolan is for sore eyes, the one at Ballyquinbeg is for sore bones, and the well at Saint Senan's is for headaches, and there is blackberry and hazelnut and plum and wild apple among and between everything.

For all its splendid, flourishing, and elaborately interconnected profusions of life, the earth is also a tomb, and the dead breathe their stories out of the ground. But those very stories, all over the world, are vanishing just as certainly as all those birds, languages, turtles, songs, and apples. They are also vanishing just as quickly.

The problem with relying on such an old and "slow" technology as a book to write about all these extinctions is not the scale of the phenomenon. It's the pace.

When the Dalcassian tribesmen built that watchtower in the Castle Field by the farm at Coolreagh, there were about 350 million people on earth. Ten centuries later, the human population had grown fourfold, to about 1.4 billion. A mere century after that, it had quadrupled again. The amount of methane in the atmosphere has more than doubled from the time of the Dalcassians. The amount of carbon dioxide in the atmosphere has grown by one-third from the time when the farmhouse was built. Global climate patterns are changing faster than in any period since the ice age began to wane 18,000 years ago. All of the warmest years since the time of Christ occurred after 1990, and suddenly we are taking all to ourselves almost 40 percent of the earth's primary productivity in the plants that we eat, the plants we feed to the animals we eat, and the forests we raze to build our cities and our homes.

The rate of extinctions in the realm of animals and plants is believed to have accelerated to perhaps a thousand times the "normal" background rate. It's hard to keep up with everything we're losing. One creature I was going to write about in this book, for instance, was a bird known as the po`ouli—a gorgeous little Hawaiian honeycreeper, confined to the upper slopes of the Haleakala volcano. But I was too late. The last one known to exist died in a cage on November 26, 2004.

So, slowing down and going for a long walk was the method I chose, and down one of the roads I walked in the East Clare hills is the Raheen Wood, where there is a tree called the Brian Boru Oak. The story told there is that the great warrior Brian Boru himself planted the oak a thousand years ago. It is a giant of a thing, living and breathing and rising into the sky out of a tangle of ferns and woodrush. They say it is the oldest tree in Ireland. It is certainly the oldest tree that remains of the Sudaine forest, which once lay thick and heavy on the Slieve Aughty mountains above Coolreagh. The forest has come to abide only in small places like the Raheen Wood because herdsmen had cut away at it for fuel and pasture, then the British cut away at it for barrel staves and ships' masts, and then came Oliver Cromwell's terror, in the seventeenth century. The forests were felled to flush the wolves and also the bands of rebels always hiding within, with their pikes and their gibberish. The last wolf in Ireland was hunted down and killed in 1786. The rebels were put to the sword. The trees kept falling.

But by then, the Irish were growing potatoes, and potato-farming served the same function as the "green revolution" in the Third World during the mid-twentieth century. It increased crop yields without addressing the root causes of hunger, such as population growth, dispossession, and the various pathologies associated with vast inequities in wealth and power. The potato had come from the New World, and it was a miracle food, just as Monsanto corn was to American industrial farmers in the late twentieth century. Then, one evening in the autumn of 1845, a strange mist settled on an Irish potato field. By the morning, the field was as black and dead as though locusts had fallen upon it. Within the space of a week, all the potatoes in Ireland had gone rotten and putrid in the ground.

Along the way to Tuamgraney, my cousin Christine and I lingered awhile in a field called An Casaoireach. Down through the years, the local people had planted yew trees in the field to keep the cattle away from all the sorrow in it. Lately, they've been planting other kinds of things, more local, distinct, and endemic varieties. But you still make the sign of the cross when you walk past, because in the ground under the trees are the bones of at least 12,000 people. There is a huge gruel tureen in the field. It's a giant iron pot that came from the Scarriff workhouse, a place that contributed many corpses to the field. An Casaoireach can be roughly translated as "Throw Them Back." Most people just call it the Famine Field. It comes from the time of An Gorta Mór, the Great Hunger.

Within three years of the potato blight that came in 1845, it was as though the heart of Ireland had been struck by an asteroid. Refugees were streaming away in every direction. When it was all over, more than a million people, perhaps as many as two million, had starved to death. By the end of the nineteenth century, Ireland had lost two-thirds of its people, and only about 15 percent of those who remained were capable of speaking their own language.

The story of extinctions today is eerily similar to the story that was unfolding in the East Clare hills in the years before the Great Hunger. It is a story of imperial capitalism, deforestation, rapid human population growth, the rise of crop monoculture, enormous disparities in the distribution of wealth, a blind faith in free trade, and the obliteration of localized culture. It's a story that always seems to lead to a field where people make the sign of the cross when they pass.

But it is only a short walk from the Famine Field to the Altar of the Sun.

Just outside the village of Tuamgraney, there are two standing stones in what once was a field, on what might be a mound, in front of Alan Sparling's house. You're welcome, then, Alan said and he shooed away his little dog. It's good for your health just to stand beside the stones, the people used to say. There are two others just like them over by Frank Hassett's gate. Or rather one, split in two, owing to someone swearing an oath on the stone and breaking the oath, many years ago.

Gráinne of the Bright Cheeks, daughter of a chief from the Slieve Aughty mountains, lived, died, and was buried. She threw herself into Lough Na Bó Girre, the Lake of the Sun, and her body floated down a little stream to a place that ended up being called Derrygraney, the Oak of Gráinne. When the people found her, they wept, and they put her in the ground.

Some things vanish from the world in an instant, and they fall among the nettles and the sorrel of those older taxonomies. Some things do not so easily pass away from the world. They move through the cartography of our deepest longings, beyond any explanation. But in all those things that we have ever killed or venerated or loved, there is something we cannot quite banish. Always, a voice. Hold fast, it says. Hold fast.

Down that deep, the fields and the stones have their own stories, and all of us, the living and the dead, the wild and the tamed, the fowls of heaven and the fishes of the sea, are a part of it now.

This is a book of those stories.

A tiger

Night of the Living Dead

I was greatly disturbed at the apparition. I walked to the left along the slope, turning my head about and peering this way and that among the straight stems of the trees. Why should a man go on all-fours and drink with his lips? Presently I heard an animal wailing again, and taking it to be the puma, I turned about and walked in a direction diametrically opposite to the sound.

—H.G. Wells, *The Island of Doctor Moreau*

The tiger lay in a clearing in the twilight, on the other side of a trickling stream, perhaps ten metres from where I stood. The cries of nightjars drifted through the trees and mingled with the thrumming of frogs and the chirping of crickets in the muggy jungle air. It was long past sunset. I'd just strolled a half-kilometre down a jungle path and across a swaying footbridge over a ravine. Then I saw it. The man-eater of English schoolboy nightmares, the great Terror of Batavia.

The tiger turned its head abruptly and glared at me. Suddenly, the crickets fell silent, and the stream with its little waterfall fell silent, and it was suddenly empty of water. Maybe somebody, somewhere, had thrown a switch by mistake. Whatever had happened, a light briefly flickered, illuminating the enclosure where I stood. I noticed a plaque—Malayan Tiger Viewing Shelter, Adopted by Chemical Industries (Far East Ltd.). There was thick plate glass separating the tiger from the outside world. There was a sign that read, Please Don't Knock, and another that pleaded, No Flash Photography, Please.

As the light flickered on and off, my view of the tiger was obscured within the image of my own face on the glass wall. At that instant, the question posed by the Romantic poet William Blake—*Tyger! Tyger! burning bright / In the forests of the night, / What immortal hand or eye / Could frame thy fearful symmetry?*—seemed to have found something of an answer in the way the innovative modern poet E.E. Cummings described the experience of seeing such animals in captivity: It's not animals that we see. Instead, it is "a concatenation of differently functioning and variously labelled mirrors, all of which are *alive*.... No mere spectacle of monsters, however extraordinary, could so move us. The truth is not that we see monsters, but that we *are* monsters."

In the fearful symmetry of Singapore's Night Safari, an auxiliary function of the Singapore Zoological Gardens, there are no cages. There are 68 lush jungle hectares, surrounded by the calm waters of the Seletar Reservoir, and the most imaginative and elaborately cunning landscape architecture is put to the work of maintaining all the illusions necessary to the suspension of disbelief. The night air is fragrant with orange blossoms and pigeon orchids. A Gir lion prowls among gaharu trees. There are banded palm civets, giant ant-eaters and babirusas—the "deer pigs" of Sulawesi's rainforest, and lesser mouse deer—the smallest of all hoofed creatures. Among the staghorn ferns and meranti trees are rare sloth bears and Malayan tapirs, those odd little things that are distantly related to both horses and rhinos and have the same colour markings as pandas.

The place is like a seventeenth-century *Wunderkammer* of the rare, the peculiar, and the vanishing, built on a massive scale. There is even an electric tram that you can take for an excursion through it all. It takes about 45 minutes.

A few kilometres away, at Singapore's famous Jurong Bird Park, you can take a monorail that will deliver you at such outlandish simulated-reality settings as the world's biggest artificial waterfall. It tumbles down the face of a 33-metre cliff at the rate of 8300 litres of water per minute, becomes a stream meandering through the world's largest "walk-in" aviary, and then gets pumped back up to the top, where it starts all over again. There are black-capped lories from New Guinea, African red-throated bee-eaters, hyacinth macaws from Brazil, Bali mynahs, Humboldt penguins, and 500 parrots from more than 100 species, almost one-third of all the parrot species on earth. Jurong houses the world's largest collection of hornbills and toucans, including the southern pied hornbill, the black hornbill, and the Great Indian hornbill.

You can wander through a series of micro-habitats taken from African savannahs, semi-deserts, and rainforests. More than 10,000 specimens of plants from 125 species create these illusions, aided wherever necessary by murals. You can walk across a swaying suspension bridge through an artificial jungle while more than 1000 Australian lories flutter around you. There are ostriches, rheas, emus, and cassowaries. The naturalistic settings and stage-light manipulations even manage to fool the birds. In the World of Darkness birdhouse, the lighting system tricks the night herons and the snowy owls and other nocturnal birds into thinking day is night and night is day. You can visit during the daytime and stroll down what looks and feels just like a starlit jungle trail. A giant mango tree grows up out of the middle of everything, and you have to read the plaque at the base of it to know that it is really just a replica of an "actual tree" in Selangor, West Malaysia.

The designers and animateurs of the Night Safari and the Jurong Bird Park have succeeded in creating a spectacularly weird simulacrum of the real world. But certain things are not so easily concealed by all those ingenious sightline considerations, psychological barriers, hidden moats, floral assemblages, and ecologically correct reproductions of landscape. There is captivity, and there is freedom. There is also what we want to believe about nature and about ourselves in that great "concatenation of differently functioning and variously labelled mirrors."

Over the course of the twentieth century, the world's tiger population fell from roughly 100,000 to about 7000. The Malayan tiger lying in the clearing in the Night Safari is a member of a species reduced to perhaps a few hundred animals, cowering in the ruins of their ancient haunts on the Malay Peninsula. The last Caspian tiger was shot in Turkey in 1970. The last Javan tiger was spotted during the 1970s in Java's Meru Betiri National Park. The last sighting of a Bali tiger, and it was a questionable sighting, was in 1976.

Most of the world's remaining tigers are Bengal tigers that survive precariously within the shallow recesses of India's national parks and wildlife reserves. A few hundred Amur tigers remain, but most of them are in zoos. There are still several hundred Indo-Chinese tigers, and perhaps 400 Sumatran tigers, but the South Chinese tiger, widely believed to be the ancestor of all tigers, has been reduced to fewer than 50 known animals, all of them in zoos. The South Chinese tiger numbered about 5000 as recently as the 1950s, before the Chinese government embarked on a pest-eradication program. The last time one was seen in the wild was in 1979. It was killed.

The 1990s began with only 14,000 Sumatran orangutans in the world; the decade ended with about 7000. In 2004, the International Union for the Conservation of Nature (IUCN) reported that the population had been halved again. This left the species officially listed as critically endangered, with populations expected to keep dropping as a consequence of forest clear-cutting and hunting. At the Singapore Zoo, you can pay to have tea with a Sumatran orangutan, all by yourself, for the equivalent of $95 U.S.

If you come to the zoo in the afternoon, you can watch pygmy hippos from behind the glass of an underwater viewing station. They loll and splash around in what for all the world appears to be a deep marsh somewhere in Sierra Leone, but nobody can say whether any pygmy hippos remain in Sierra Leone. There are only about 7000 left in all of West Africa, and most of them are confined to Liberia's Sapo National Forest, but their numbers are declining rapidly. Logging companies are turning their swamp forests into wastelands, and the hippos are being killed for food and by trophy hunters who want their teeth. With the collapse of order in that part of the world, the civil wars and insurrections, the hippos' prospects aren't good. A related group of pygmy hippos was once common in Nigeria; in 1969 they were found to be a distinct subspecies, but there have been no confirmed sightings since then.

Another resident of the Singapore Zoo is the douc langur, an extravagantly coloured little monkey that suffered enormously during the American defoliant-bombing of Vietnam. Douc langurs are found only in Vietnam and neighbouring Laos. They are being hunted for food, for the pet trade, and for the folk-medicine market. Their forests are falling to chainsaws. Another zoo inmate is the endangered and increasingly rare ruffed lemur from Madagascar. Known in its home range by a name that translates as "night-wandering ghost," the ruffed lemur is one of the world's most unobtrusive primates, quietly going about its nocturnal rounds, barking only to warn its comrades of danger. It is a key pollinator for several plant species because it has an inordinate desire for nectar, and tends to go from flower to flower, its nose covered in pollen. There are proboscis monkeys at the zoo, too. They're the weird-looking, big-nosed monkeys from Borneo. Their numbers are dropping sharply because of the spread of timber operations and the rise of oil-palm plantations. The lion that strolls among the gaharu trees is from a vanishing population: only 200 of those regal creatures remain in India's Gir forests.

The hyacinth macaws at Jurong are critically endangered, numbering fewer than 300 in their home forests in Brazil. The Bali mynahs at the bird park are among the rarest of the world's birds; only a few dozen persist on their home island, outnumbered more than ten to one by the Bali mynahs in the zoos and aviaries of the world. Humboldt penguins too are undergoing a precipitous decline in their home waters, in the Pacific, partly because the fish they eat are being depleted by fishermen, in whose nets the penguins also often perish, and partly because their nesting sites are being mined for guano. Great Indian hornbills are disappearing, too, because the forests of India are disappearing, but also partly because the birds have long been a favourite of bird collectors. They're the largest of the world's hornbills, standing well more than a metre in height, and they're possessed of such endearing habits as offering their captors morsels of food. For decades, one of the London Zoo's most popular animals was Josephine, a Great Indian hornbill. She died in 1998, at the age of 52.

About 2000 animals from 250 species are held at the Singapore Zoo and the Night Safari, 9000 birds from about 600 species at the Jurong Bird Park. These institutions are routinely and deservedly praised by the appropriate international bodies as well-managed and progressive places, but there are other words that might be used to describe them. Here are three: Hospice. Necropolis. Tomb.

The Great Indian hornbill, the Bali mynah, the hyacinth macaw, the proboscis monkey, the ruffed lemur, the douc langur, the pygmy hippo, the Sumatran orangutan, and the Malayan tiger belong to a special class of rare and vanishing creatures of the wild world known as the "living dead." It's a term biologists have begun to use to describe those species that are not expected to escape extinction without significant human intervention, such as captive breeding. Among the world's endangered mammals, birds, and reptiles, already 1500 species are expected to be wholly dependent upon captive breeding by 2050.

Specifically, the term "living dead" is used to describe species that have been rendered incapable of independent survival because other species upon which they depend are disappearing or are already gone. The living dead include species that exist mainly in zoos, such as the Amur tiger, and those suffering "latent extinction," which appears to be the douc langur's condition, as it slowly withers away as a result of habitat loss. Not all the critically endangered species on earth are necessarily counted among the living dead. It's hard to say whether a species has crossed into that netherworld unless you have a pretty clear idea about its long-term prospects. A key factor to consider is the extinction debt racked up from habitat loss that has already occurred.

One grim example of the way extinction debt works its misery comes from a study published in the journal *Conservation Biology* in 1999. Its author, Guy Cowlishaw, of the London Zoological Society, looked at the extinction debt incurred from forest clearing in Benin, Burundi, Cameroon, the Ivory Coast, Kenya, and Nigeria. Cowlishaw determined that the impact of logging would likely result in the extinction of one-third of those forests' primate species. That was the outstanding debt, even if all logging and poaching stopped the moment that Cowlishaw published his findings. It might take a century for those primate species to limp along, slowly paying off the debt until reaching equilibrium with their very extinction. But the debt will be called, and it will be paid. The trees keep falling. West Africa is expected to lose 70 percent of its already diminished forests by 2040. East Africa's forest losses are projected to be as high as 95 percent.

Singapore itself provides a vivid example of the way extinction debt works, as well as a rare glimpse into the way the world's current extinction crisis might be expected to unfold in the coming years.

The island nation of Singapore is situated in the humid tropics, which is the epicentre of the planet's current crisis in the extinction of wild things. The tropics are playing a central role in the ongoing story of extinctions because tropical forests contain the world's deepest reservoirs of terrestrial species diversity, and it is in the tropics that forests and other "old-growth" ecosystems are disappearing the fastest.

It's mainly this vanishing of tropical habitat that results in estimates putting the current global extinction rate as high as a thousand times the "normal" background rate. Those estimates are extrapolated from the relationship between habitat size and species diversity and a calculation of what habitat loss will mean for species loss. Another method involves tracking the progression of species through their trajectories on the status lists maintained by the IUCN, the main international body that monitors the collapse of biological diversity. These methods may seem a bit speculative around the edges, but they tend to be confirmed by the hard data produced by specific analyses of trends in well-known families of birds, plants, and animals in well-defined locales. Singapore is precisely one such well-defined locale. And unlike much of the tropical world, Singapore is positively robust in empirical data related to biological diversity and its withering.

Avocational naturalists and birdwatchers have been going about their business in Singapore since the earliest times, compiling meticulous records of the local flora and fauna. The island already had its own formal naturalists' society in the 1950s. It has been a tireless little group, providing at least a marginally effective voice for conservation despite being often only barely tolerated by the authoritarian regime that has controlled the country since the 1960s. But Singapore's naturalist traditions reach all the way back to the founder of the former British colony, Sir Stamford Raffles. Although Raffles earned his reputation as a vigorous but fair colonial administrator and an able challenger of Dutch commercial interests in the East Indies, he was also an avid collector of animal and plant specimens. That is what they will tell you at Singapore's Raffles Museum of Biodiversity Research, where the Old Man's tradition has been kept alive in collections, which include 18,000 plant specimens (2000 strains of fungi, even) and the carefully preserved bits of 300,000 dead animals from more than 10,000 species. The place is a marvel.

Biologists Barry Brook, Navjot Sodhi, and Peter Ng, an Australian and two Singaporeans, reckoned that by looking at what had happened to biological diversity in Singapore, they might get a better grasp of the real impact of habitat loss elsewhere in the tropics. Mindful that the island is especially bollixed from an ecological point of view, Brook, Sodhi, and Ng reckoned they might draw from Singapore's experience some well-informed projections about the fate of biological diversity in the years to come and test those global trends in extinction that are otherwise unavoidably inferred from statistical models or by extrapolation. The Singapore study was published in the journal *Nature* in 2003.

The study revealed that fully half of the island's species existed only as relics within the mere one-quarter of 1 percent of the land mass that had been protected as forest reserve. Many of those species carried on in the weird half-life of the extinction debt arising from past habitat loss, a debt conventionally paid with the eventual oblivion of species. Three such Singaporean animals found to be among the living dead were the white-bellied woodpecker, the banded leaf monkey, and the cream-coloured giant squirrel. There were only four of the woodpeckers left, fewer than 15 of the monkeys, and fewer than 10 giant squirrels. Singapore had lost at least 95 percent of its forest cover since Raffles's time. Documented and conservatively inferred extinctions had occurred among 80 percent of the island's fish species, almost 80 percent of its mammal species, more than 70 percent of its plant species, about 60 percent of its bird species, 70 percent of its once-abundant butterfly species, and 70 percent of its amphibians.

Brook, Sodhi, and Ng calibrated these rates of local-population losses against the patterns of deforestation throughout Southeast Asia, which are projected to result in the disappearance of 74 percent of the region's forests. They concluded that somewhere between 13 and 42 percent of all Southeast Asian species—mammals, birds, plants, amphibians, decapods, phasmids, butterflies, reptiles, the lot—were more or less done for. Furthermore, half of the region-wide extirpations that would follow from forest loss could be expected to result in global extinction. That's because so many of Southeast Asia's life forms are endemic, which is to say they occur only locally.

At the Singapore Zoo, oblivious to the world outside, the living dead carry on. The proboscis monkeys are producing offspring. The zoo boasts the highest numbers of orangutans bred in captivity at any one institution—21. Twenty-two Malayan tigers have been born at the Singapore Zoo since 1973, along with 28 chimpanzees and 3 douc langurs, those exceedingly rare monkeys from the defoliated mountains of Vietnam and Laos. Other captive-bred members of living-dead species at the zoo are golden lion tamarins and white rhinos. Fourteen pygmy hippos have been born there. Around the world, 178 pygmy hippos live in 74 collections, and most of those hippos were born in zoos, to zoo-born parents. At the Jurong Bird Park, meanwhile, captive-born offspring have been hatched among more than 100 bird species, including many endangered species. Jurong is the only institution in the world to have successfully hatched fledglings from the southern pied hornbill, the black hornbill, and the Great Indian hornbill.

Among the growing ranks of the living dead, the ancient paradigm of evolution, as Darwin described it, is over. If their kind are among us at all a century from now, they will be wholly different from the creatures humans first encountered. They will not be "wild" animals at all. They will be functions of artificial selection. They will live on in zoos, and perhaps some large parks. They will live in a world populated by animals we have chosen, with traits we have chosen, and in numbers we have chosen. If they live on in wilderness at all, it will be a wilderness of our own making. They will live in a simulacrum of the real world, in places like the Jurong Bird Park or the Singapore Zoo and its adjacent Night Safari grounds.

Every year, the Singapore Zoo attracts 1.5 million visitors. Ah Meng, the Sumatran orangutan with whom you can pay to take tea, received a special award from the Singapore Tourism Board in 1992. She had raised five of her own babies at the Singapore Zoo.

She became a grandmother there.

In Greek myth, the chimera was a grotesque fire-breathing animal with a lion's head, a goat's body, and a dragon's tail. The term has come to mean a fanciful creature made up of different animals, and Singapore actually has its own chimera. "The Merlion" is an absurd-looking creature with the head of a lion and the tail of a fish. According to the official story, the Merlion comes from a thirteenth-century Malay legend and Singapore chose to adopt it as its national symbol. In the real world, the capitalist-authoritarian regime of Lee Kuan Yew, the Singaporean Tito, invented the Merlion in the 1960s, around the time it was jailing newspaper editors and left-wing intellectuals and shuttering Nanyang University. You will read in countless travel articles that the Merlion chimera has been considered the island's guardian since ancient times. Actually, it wasn't until the 1990s that tourism bureaucrats put their final touches on the fabrication of the legend. In Singapore, you can't even use the symbol of the Merlion for anything without government approval.

Singapore lies just off the southern tip of the Malay Peninsula, a few sea miles from several small islands immediately adjacent to the coast of nearby Sumatra. For all its grasping materialism and stifling political culture, Singapore has remained secular, multicultural, and stable, and has avoided the violence and despotism that convulsed nearby Malaysia and Indonesia throughout the latter half of the twentieth century. But four million people are packed into a country smaller than the Canadian city of Edmonton. As the 2003 *Nature* study showed, Singapore has almost completely devoured what was a largely uninhabited tropical island when Stamford Raffles and the British East India Company arrived in 1819. But the government has not been satiated by the island's living things: it has been pulling down the mountains and bulldozing them into the sea to make more room for itself in the shallows of the Singapore Strait.

With its cheek-by-jowl high-rise office towers, its vulgar residential complexes, its surfeit of American fast food franchises, multinational corporation branch plants, and off-the-shelf urban architecture, Singapore has been subsumed within the dreary homogeneity descending upon the cultures of the world. By the end of the 1990s, most of Singapore's buildings were less than 30 years old. The city is a bit difficult to describe, in fact, because it is so much like everywhere else. Even its oldest and largest graveyard, the venerable Bidari Cemetery, was bulldozed to make more room for urban development. In 2004, William Lim, a leading authority on urban development and architecture in Southeast Asia, described Singapore as a place where urban planners have "systematically removed and destroyed unprotected city areas and historical sites that had acted as containers of history, values and cultures."

During my time there, I found regular solace at the In The Name Of Allah The Most Gracious The Most Merciful Mohd Rajeen & Brothers Café, a happy little establishment in Arab Street. The Muslim quarter is one of the few places left in Singapore that feels and behaves like a real place. Apart from a handful of neighbourhoods that have somehow retained the old Malay *kampong* village atmosphere, Little India and Chinatown are the only other significant districts where some authentic local sensibility can be found. The rest of the city has been adequately described by the celebrated Dutch architect Rem Koolhaas as a place devised by "pure intention: if there is chaos, it is authored chaos; if it is ugly, it is designed ugliness; if it is absurd, it is willed absurdity. Singapore represents a unique ecology of the contemporary."

Singapore more or less straddles the intersection of the "security first" and "markets first" roads the UNEP scientists were talking about in the *Global Outlook* report I referred to in the Prologue. Its people, most of whom are ethnically Chinese, have suffered the country's Orwellian conditions of official truth, reinvention, construction, demolition, and reconstruction with a remarkably jolly grace. They take what little comfort they can in the country's strides toward liberty and democracy. In 2002, for instance, the government relaxed its laws against the importation of chewing gum, but the people still have to be careful about asking too loudly for much more. As Singaporean journalist Cherrian George reports in his brave book *Singapore, The Air-Conditioned Nation,* the September 1, 2000, inauguration of a speaker's corner in Ho Lim Park marked the first time since the colonial era that Singaporeans were allowed to address their fellow citizens publicly without a government licence. Still, having sustained 40 years of single-party rule by the People's Action Party (PAP) since independence, Singapore is burdened by press-freedom restrictions that led Reporters Without Borders to rate it only slightly above North Korea and Myanmar.

The rating upset Information Minister Lee Boon Yang, who explained that in Singapore, journalists were expected as a matter of course to contribute to "nation building." By this phrase, the minister was referring to the government's obsessive preoccupation with deforestation, urban expansion, construction, population growth, the enforcement of official mythology, and the burial of history—in other words, all those things contributing to the extinction of animals throughout the tropics, and to the extinction of local cultures, distinct urban landscapes, and ways of life. Not satisfied with having the third-highest population density of any country after Hong Kong and Monaco, PAP officials unveiled a plan in 2004 to boost population growth rates by offering multi-children families tax breaks, reduced maid fees, better maternity benefits, cheap mortgages, and heftier family-allowance packets.

The entire country, which has practically no natural resources of its own, is the product of the limited company, which evolved into the multinational corporation—a statutory chimera fused with the same legal rights that Western democracies found first only in living, breathing human beings. It is well represented by the Merlion, of which there are four officially recognized statues. One of them gazes out from an imposing waterfront commercial-entertainment-office complex at One Fullerton, down at Collyer Quay. It magically spits water into the Singapore River. The most grotesque is the gigantic Merlion statue on the artificial island of Sentosa. You can even go inside it, where there are gurgling sound effects and murals depicting fanciful pirates. The Sentosa Merlion is 37 metres tall. It emits laser beams from its eyes and smoke from its arse.

By the first decade of the twenty-first century, there were many chimeras moving through the world. The final walls between nature and artifice and between captivity and freedom are being scaled and breached at every rampart. Less than a century after H.G. Wells wrote *The Island of Doctor Moreau,* transgenic laboratory chickens were trilling like quails and a Canadian company had spliced the genes of a spider into the genes of a goat in the hopes of processing from the goat's milk a "biosteel" with the tensile strength and flexibility of a spider's web. A transgenic zebra fish, the "glofish," has been genetically engineered by a Texas company to glow fluorescent red in the darkness of an aquarium tank. In 2004, a California firm, saucily named Genetic Savings and Clone, went into the business of cloning pets at $50,000 a clone.

Even that thick pane of glass between the tiger and the rest of us is starting to crack. Scientists have bred mice with human brain cells, pigs with human blood in their veins, and sheep with human cells growing in their hearts. In 1987, the U.S. Patent and Trademark Office, following upon a U.S. Supreme Court ruling, declared that man-made organisms could be patented. Canadian and European government agencies eventually followed the American lead. Stem-cell research has opened up a vast potential for eliminating diseases and genetic abnormalities, and there are few rules to go by.

As early as 1988, the University of Virginia's Joseph Fletcher, a founder of the field of study known as bioethics, spoke approvingly of genetic manipulation to create "parahumans" with the physical capacity to do dangerous and demeaning work without the burdensome encumbrance of human rights. In 1997, two Americans, biologist Stuart Newman and biotech critic Jeremy Rifkin, applied to patent a "humanzee," a speculative part human, part chimpanzee species. Their application was a provocative stunt intended to force debate, and Newman and Rifkin were turned down. But by November 2002, U.S. and Canadian scientists had gathered in a closed meeting in New York, co-sponsored by Rockefeller University and the New York Academy of Science, to discuss the feasibility of creating a human–mouse chimera. The discussion centred on the possibility of injecting human embryonic stem cells into an early mouse embryo, to test whether it was possible to create a "mouse" carrying the full complement of human genes.

Freeman Dyson, with the Institute for Advanced Study in Princeton, New Jersey, argues that we should all be pleased with the possibilities. "Now, after some three billion years, the Darwinian era is over," Dyson wrote in the Massachusetts Institute of Technology's online *Technology Review* magazine in March 2005. "Cultural evolution is running a thousand times faster than Darwinian evolution, taking us into a new era of cultural interdependence that we call globalization. And now, in the last 30 years, *Homo sapiens* has revived the ancient pre-Darwinian practice of horizontal gene transfer, moving genes easily from microbes to plants and animals, blurring the boundaries between species." Dyson is positively ecstatic about the prospects: "There will be do-it-yourself kits for gardeners, who will use gene transfer to breed new varieties of roses and orchids. Also, biotech games for children, played with real eggs and seeds rather than with images on a screen. Genetic engineering, once it gets into the hands of the general public, will give us an explosion of biodiversity."

But a rather less utopian future seems just as likely, a future a bit too much like the hideous island world H.G. Wells imagined in the late nineteenth century. "Of course these creatures did not decline into such beasts as the reader has seen in zoological gardens—into ordinary bears, wolves, tigers, oxen, swine, and apes. There was still something strange about each," the narrator relates in *The Island of Doctor Moreau*. "In each, Moreau had blended this animal with that. One perhaps was ursine chiefly, another feline chiefly, another bovine chiefly.... The dwindling shreds of the humanity still startled me every now and then—a momentary recrudescence of speech perhaps, an unexpected dexterity of the fore-feet, a pitiful attempt to walk erect."

There were no creatures quite like this on hand at the Night Safari in Singapore. There were, however, many representative specimens of the new, globalized, and "culturally interdependent" *Homo sapiens* that Freeman Dyson would approve of. They came in the form of affluent and disaffected young Singaporeans roaming in half-drunk flocks, their skulls fitted with wires from their iPod headsets, all accompanied by the sound of ceaselessly chiming cellphones. They were draped in the predictable array of Gap knock-off sweaters, brand-logo sweatshirts and Britney Spears T-shirts. This sort of thing is harmless enough, but it is not unrelated to those things William Lim observed about the deracinated, dehistoricized place Singapore had become. It is directly related to those things the Singapore Zoo's animateurs would rather we not notice. It is about the bleeding away of differences in the living world, and of differences between captivity and freedom, between the real and the fabricated, between what we would want to believe and what is really happening, all around us.

Linda MacDonald Glenn, senior fellow with the American Medical Association's Institute for Ethics, says that such differences are becoming obscured along with the differences between species, and between humans and animals. On the immediate horizon is a world of wholly new creatures, forged from the DNA of different species. It is a world of cyborgs, parahumans, and Jurassic Park revivals of extinct animals. With exogenous pregnancies, the development of implantable brain chips, and transgenic engineering, we are already on the frontier of that world. By these innovations, we are "forging the next step in our own evolution," Glenn says. "Future developments will likely challenge our concepts of what it means to be human," she predicts. "What once was fiction has now become fact."

But certain things have not changed and are not likely to. "Nature" has always been both an objective reality as well as the sum of our ideas about it. At the same time, nature is always changing, and so are the ways we think about nature. In the "wild" world, it has never been easy to separate fact from fiction, as the story of the Malayan tiger reveals.

The tiger known to the Malay people was not a "wild" animal in the modern understanding of the term. It was not the same animal out of those ripping tales in *Boys of the British Empire* magazine during the 1920s and the *London Illustrated News* in the 1930s. When the Dutch and the British first set out to trade with, colonize, and ultimately conquer the East Indies, no terror of the dark jungle loomed more ominously in the colonial imagination than the tiger. Even now, to Westerners, nothing better typifies the "wild" world—the "natural" world—than the proud, savage, and noble tiger. But the Malay peoples never saw the tiger quite this way.

The Malay relationship with tigers perplexed outsiders from the earliest times. In a fifteenth-century Chinese account of a visit by Chinese traders to a Malay village, tigers occasionally turned into men, becoming "weretigers" that walked undetected among people. In the late seventeenth century, the Scottish trader Alexander Hamilton insisted that people in the vicinity of Malacca enjoyed the company of tigers and rode around on their backs. There are several early accounts, not necessarily fanciful, of Malay people befriending tigers and sharing food with them, and of tigers sharing their kills with people.

A widespread Malay custom that persisted well into the late nineteenth century was the tradition of tribute to a *macan bumi,* the "village tiger." People took turns leaving meals of goats or chickens at a sort of shrine on the village outskirts, and the offerings were regarded as a kind of tax the tiger levied. European colonial administrators were at once baffled and disturbed by the *macan bumi* custom. They regarded it as a dangerously foolish superstition that also inhibited the march of progress and the clearing of ferocious and noxious beasts from the jungles. But Sir George Maxwell, writing around 1900, reported that in some Malay villages, even small children were adequate to the task of driving off a tiger that strayed suspiciously close to a herd of cattle.

In the book that resulted from his expansive, ten-year investigation of the place tigers occupied in the Malay consciousness, *Frontiers of Fear: Tigers and People in the Malay World, 1600–1950,* Peter Boomgaard suggests that there was a lot more to the *macan bumi* tribute than mere backwoods mumbo-jumbo. A tiger habituated into the role of a *macan bumi* was far less likely to carry off villagers or cattle. A village tiger was also understood to drive out other, unfamiliar tigers, especially saucy young males looking to establish their own home territories. That was the village tiger's part of the bargain, and should a tiger fail in its duties, and a new tiger showed up at a village, the local dignitaries would beseech the new animal to go away. A commentary on just such an event, observed on Sumatra, comes from no less a personage than Sophia Raffles, Sir Stamford Raffles's second wife: "When a tiger enters a village, the foolish people frequently prepare rice and fruits, and placing them at the entrance as an offering to the animal, conceive that, by giving him this hospitable reception, he will be pleased with their attention, and pass on without doing them harm." She does not report whether the villagers' entreaties had any effect.

But there are few generalizations that can be made about the Malay attitude toward tigers. Some tribes called him "grandfather." Others would not utter the word for "tiger" at all, but would quietly make the sign of a claw, with their hands, reportedly for fear of arousing his attention. In Java, tigers were said to make annual pilgrimages to the "heathen" pre-Muslim shrines at Arca Domas. Whenever Malays attempted to explain their ideas to Europeans, the baffled inquirers invariably reached the conclusion that to the Malays, tigers were the embodiment of long-dead ancestors or the physical manifestations of wandering human souls. The anthropologist Ivor Evans was driven to distraction trying to sort it out: "For all I know, tigers may be thought to be human beings who have assumed an animal shape."

In some districts, it was only when an individual tiger persisted in killing villagers or their livestock that the locals took matters into their own hands, and even then the hunters took great pains to beg the animal's forgiveness before killing it. In other cases, the hunt was an elaborate ritual overseen by a tiger charmer. After the animal was trapped, a tiger charmer would spend days explaining in great detail the gravity of its transgression, why it had been wrong to kill the village livestock, and why the villagers had been left with no recourse but slaughter. In still other districts, the tigers themselves were regarded as enforcers of village custom, giving rise to the assumption that anyone who was eaten by a tiger must have been a thief or an adulterer or had secretly recited a taboo poem, but in any case must have had it coming.

Some Sumatrans would not kill a tiger no matter how grievous its offence. Some Javans wouldn't hesitate to kill a tiger that had provoked a village by killing a cow. Colonial officials routinely complained that regardless of the inducements offered, a Malay could rarely be prevailed upon to kill a tiger. One headman of a Sumatran village where a tiger had been reluctantly killed refused the generous bounty available to him on the grounds that taking money for the act would be like selling an ancestor.

Malayan tigers were actually quite rare before the eighteenth century. But as pasture land and plantations began to open the unbroken forests, making more room for boar, deer, and other ground-dwelling herbivores and omnivores, such as goats and cattle, tiger populations rose. Imperialist expansion, the adoption of new technologies, and a shift to more permanent towns and an agrarian economy were changing things radically. The ways Malay people thought about tigers changed too. By the early nineteenth century, an average of 500 Javan people were being killed by tigers every year. During the same period, tigers killed roughly 1000 people a year on Sumatra. In the Sumatran district of Lampung, in 1820, at least 675 people were reported to have been killed in attacks. A kind of order had collapsed, and the Malay peoples responded in different ways. In some places, to the great delight and encouragement of Dutch colonial administrators, the ritual slaughter of captured tigers, in gruesome public displays, became commonplace. In other places, the old rules held.

Now, like then, it is hard to tell where the old rules will hold and where they might give way. Nature changes and evolves, just as our ideas about nature change and evolve, sometimes because of scientific discoveries and sometimes because of changes humanity has forced upon nature. Even the Malayan tiger, a showcase among the nocturnal animals in Singapore's Night Safari, was not always exclusively an animal of the night. It's almost certain that the tigers' habit of hunting at night is largely a recent response to the ubiquity of humans, who usually go about their affairs during daylight hours. If that is so, it is an adaptive mechanism the tigers figured out on their own, as a way to share the world with us.

Human beings change nature. In its turn, nature forces changes upon us, upon the way we imagine nature and what we "know" about nature. And round it goes. It's the way the dialectic has always worked. What's different now is the pace and scale of the dialectic, which is being driven partly by the pace and scale of extinctions and partly by the pace and scale of what we are coming to know about extinctions.

What we know is that the earth is undergoing a period of mass extinction comparable to only five cataclysms that have occurred over the past 440 million years. For crude shorthand purposes, we'll call those other extinctions the Ordovician extinction, the Devonian, the Permian (the most destructive of them all), the Triassic, and the Cretaceous (the one that carried off the dinosaurs). We also know with some certainty, from the fossil record, that the earth's current wealth of biological diversity may well be twice the average of any period in the history of life on earth. Given the accelerating rate of extinctions, there is little comfort to be taken in this knowledge.

One big unknown is just how many species of life there are. Reckoning those numbers can be contentious, only partly because of disputes and changes in our understanding of what constitutes a species, what constitutes a subspecies, and so on. While I was writing this book, for instance, the Malayan tiger was turning out to be not nearly as closely related to the Indo-Chinese tiger as once thought. It was being transformed by the International Species Information System (ISIS) in a way that moved it out of the subspecies *Panthera tigris corbetti* into its own subspecies, *Panthera tigris jacksoni* (after tiger conservationist Peter Jackson).

Another problem with definitively stating just how much diversity is out there is the sheer magnitude of the task. There are fewer than two million known species, and those figures are generally held to represent only a small fraction, maybe 10 percent, of the species in the world. About 15,000 "new" species introduce themselves to science every year. These additions to the known world include a mélange of well-known species that finally end up formally recognized and named. They also include completely "new" orders of life found in such places as deep-sea thermal vents, as well as discoveries arising from innovations in microscopy, revisions in our understanding of species' compositions following advances in analysis of mitochondrial DNA sequencing, and so on.

Only a small fraction of these additions to the sum of the known world come from species that people have just encountered. Such discoveries are becoming more commonplace as the earth's last tracts of wilderness give up their secrets to humanity, yet even big things in well-explored areas can hide from taxonomists for a long, long time. North America's largest land tortoise, for instance, was unknown to science until 1958. The 14-kilogram animal was found in Mexico's Chihuahuan Desert; it didn't take long to earn an endangered classification. Another "new" species is the Bornean cat. The outside world had been dimly aware of it, from a few nineteenth-century skins and skulls, but in 1992 scientists confirmed its existence through genetic analysis of a female that had been captured and photographed in Sarawak. In the picture, the cat's face looks like something that would result from crossing a cougar with a monkey. But it was probably just an unflattering portrait.

In addition to the great revelations that deep-sea submersible technologies are bringing back from the depths, the oceans continue to provide surprises. Several rare beaked whales are known to exist out there, or at least to have existed, although with the lack of concrete evidence it's not easy to describe them as ever having been really "discovered." The Indo-Pacific beaked whale is known from only two skulls, one found on an Australian beach in 1822 and another in Somalia in 1955. The pygmy beaked whale is the smallest (four metres in length, tops) and most recently discovered. It wasn't formally described by science until 1991, from the evidence of only a few sightings and a handful of dead whales found by Peruvian fishermen. Another, the Andrews' beaked whale, is known only from carcasses washed up on the beaches of South Pacific islands.

The "new entries" in-basket for species is absolutely crawling with bugs. The great entomologist E.O. Wilson, who looms over the study of biological diversity like some latter-day Darwin, discovered a new ant in 1989 in the Washington, DC, offices of Kathryn Fuller, president of the World Wildlife Fund's American chapter. Bug discoveries happen all the time.

Then there are the special cases that confound our understanding, such as "wild" creatures that exist in captivity but whose origins are a mystery. In a shrine at Chittagong, in Bangladesh, black, soft-shelled, three-clawed turtles have subsisted on the offerings of pilgrims for about 1200 years. Five hundred of these animals remain. Nobody knows where they came from, but pilgrims insist they are the incarnations of the first followers of the saint to whom the shrine is dedicated.

Every now and then, cheerful news surfaces about the "rediscovery" of a species that everybody thought was extinct. The aye-aye of Madagascar, a nocturnal primate that looks like a cartoon goblin, was routinely killed by local people, who considered it a portent of evil. Its numbers had dwindled to the point that it seemed almost certain to have vanished entirely from its remnant home forests, but then, in 1986, two adults were found and captured. The Jamaican iguana was written off as extinct in 1946 but showed up again in 1990. A captive-breeding program has produced 100 of the animals at Jamaica's Hope Zoo, but hopes for their reintroduction are not high, due to the depredations of dogs. Jerdon's courser, a long-legged, plover-like bird discovered in India's Andhra Pradesh state in 1848, showed itself in public only three times during the nineteenth century. The Bombay Natural History Society mounted a faint-hope search for the bird and found a few, in 1976, in Telugu. In 1993, an American scientist came upon a Madagascar serpent eagle that had been thought extinct in the 1930s. And in a most dramatic find, Spanish scientists looking for reptiles in the Canary Islands in 1999 came upon six specimens of a type of giant lizard, on a cliff on the island of La Gomera, that had been believed extinct for 500 years.

All this sort of knowledge is fairly easy to take in. We can always say, well, if there's a species out there, we can name it, and if it disappears from the world, it's extinct. But progress, as it's often called, marches ahead. The alchemies of cryogenics, transgenetic manipulation, and the emerging biotechnologies that zookeepers are developing raise the very real possibility that many of the "living dead" species of the world will never become extinct, at least not in the way that geneticists or taxonomists use the term.

These same advances are raising the prospect of bringing extinct species back from the dead. In 2002, scientists associated with the Australian Museum in Sydney extracted DNA from the pickle-jar fetus of a Tasmanian wolf, a marsupial thylacine driven to extinction in the 1930s. The hope was to reassemble its genetic structure to eventually clone and produce live specimens. Then Britain's Natural History Museum embarked on an ambitious plan to preserve the genetic blueprints of all the world's endangered species, just in case future scientists proved capable of pulling off the kind of magic trick the Australians had in mind for thylacines.

Meanwhile, "taxon advisory groups" all over the world are developing species survival plans overseen by the ISIS. The scientists are routinely confronted with the most vexing questions about what to do *now,* not what they might do in the future. Which species should be simply abandoned? Which locally adapted population should we try to save? Which unique subspecies in the wild should we write off in favour of deepening the genetic reservoir available to the species as a whole? In the zoos of the world, the old business of tossing hay bales at caged elephants is rapidly giving way to the work of preserving the embryos, sperm, and tissue of endangered animals in nitrogen freezers. Newborns among many critically endangered species are as likely as not to be products of artificial insemination, test-tube fertilization, electro-ejaculation, or matches arranged by geneticists. Semen is being extracted from anesthetized tigers and stored away in frozen test tubes. The fertilized embryos of endangered gaurs are being implanted in Holstein cows. Horses are giving birth to zebras.

The faint possibility that "extinction" might not be so final after all has reopened hotly contested debates about whether parks and protected areas, zoos, aviaries, and aquariums could serve as the basis for a new "ark." The idea is utopian, based as it is on the hope that some day the flood of extinctions will recede, a dove will appear on the horizon, and we will find ourselves on the shore of some enlightened age where we can let all those animals loose again. But to get a better glimpse of the future of zoos and parks, we need to look back at where they came from and what uses they've served. And in that story, too, Singapore has played a leading role.

The story of the modern zoo begins with Sir Stamford Raffles, founder of the colony of Singapore, servant and champion of British imperialism. Raffles is well remembered in Singapore. Along with the Raffles Museum of Biodiversity Research, there is Raffles Boulevard, Raffles Road, and Stamford Road. There is the Raffles Hospital, Raffles Plaza, and Raffles Park. Most famously, there is Raffles Hotel, the opulent nineteenth-century Singapore institution associated with Rudyard Kipling, Joseph Conrad, and Somerset Maugham. I enjoyed a gin and tonic there, a few paces from the spot where the last tiger in Singapore was shot and killed in 1902 underneath a pool table in the Bar & Billiard Room.

Raffles himself introduced several species to science. His own private menagerie included orangutans, monkeys, tigers, gibbons, and bears, but the species with which he is most intimately associated is *Rafflesia arnoldi,* the largest and arguably the most hideous flower in the world. A parasite upon a particular sort of jungle vine, the Rafflesia produces a stemless flower that can reach a metre in width and looks something like a gigantic rotten orange mushroom that some child has tried to sculpt into a flower using a dull carving knife. The flower blooms for only a few days, and when it does, it emits an overpowering smell, just like a heap of rotting flesh, which attracts flies, its primary pollinator.

Raffles's more enduring legacy is in having established the London Zoological Gardens, at Regents Park, after returning to England from his tumultuous and successful colonial service. The institution laid the foundations for the modern zoo and even gave the word *zoo* to the English language.

The Regents Park zoo wasn't the world's first grand bestiary. The earliest is believed to have been a collection of perhaps several thousand animals at Saqqara, in Egypt, about 4500 years ago. Its holdings included 1134 gazelles, 1305 oryxes, and more than 1200 antelopes of a variety of species. Three thousand years later, the pharaoh Thotmees III established an even more vast menagerie and botanical garden, with monkeys and leopards. On the far side of the world, 500 years after that, the Chinese emperor Shen Nung, the Divine Husbandman, is said to have established a huge "Garden of Intelligence" containing an impressive array of animals and plants. In the early sixteenth century, when the Spaniard Hernando Cortés arrived in the Valley of Mexico, he came upon a grand zoo in Montezuma's great city, Tenochtitlán.

But what Raffles set out to assemble was the most expansive collection of animals in history. He was not content to create a grand extravagance, a mere collection of curios for the "vulgar admiration" of the hoi polloi. It would be instead a giant laboratory, an institution that would facilitate the study of the animal world. He wanted something on a scale and of a class wholly different than, say, Polito's Royal

Menagerie, a hideous place in London where lions, tapirs, monkeys, tigers, an elephant, and a rhinoceros were kept in cages so small the animals could barely stand up in them.

Raffles set out to create "a grand zoological collection in the metropolis" that would cater to fashionable company and would be to zoology what the Royal Botanic Kew Gardens was to botany. As Kew set out its collections according to the order devised by Carolus Linnaeus (the founder of modern taxonomy), the London Zoo would arrange its exhibits according to the best semblance of taxonomic propriety its zookeepers could manage. One of Raffles's chief advocates and colleagues in the zoological-garden enterprise was the great Sir Joseph Banks, who accompanied Captain James Cook on his historic explorations of the South Seas, served as the director of Kew Gardens, and presided over the venerable Royal Society. The London Zoo opened, with 200 species, in 1828, shortly after Raffles's death. In its first few years, the zoo was open only to wealthy subscribers. By the 1840s, anyone capable of paying the one-penny admission could attend.

For the animals of the world, though, the opening of the London Zoo was not a welcome development. It set off a frenzy of imitators throughout Europe and North America, all supplied by constantly moving caravans of caged animals from the most remote reaches of the empire. Singapore emerged as a major centre of the global animal trade and held that position for several decades. Wholesale markets for wild animals could be found all along Rochor Road, and Singapore provided a key base of operations for Frank Buck, one of history's most notorious animal dealers. Buck made a documentary film about himself in the 1930s, in which he claimed that over his 30-year career he had captured or bought and sold more than 60 tigers, 49 elephants, 5000 monkeys, and 100,000 birds.

Leopards, lions, kangaroos, crocodiles, bears, cougars, cheetahs, giraffes, wolverines, gorillas, orangutans, chimpanzees—a seemingly endless list of species, in seemingly endless numbers, filled the cargo manifests of countless ships. Most of the animals died along the way. They succumbed to disease, heatstroke, and exhaustion, perished in shipwrecks, and starved or died for lack of water during the long trek across the deserts of North Africa. Some shipments were written off entirely, the whole "cargo" having died en route. Of those animals that survived the journey to European zoos, most were dead within three years of arriving. In Africa, gorilla populations sustained enormous losses. As late as the mid-twentieth century, only a small fraction of the gorillas captured for the zoo trade survived their journeys, and most died within a few months of arriving in their cages. A common ailment was "indigestion," a euphemism for the consequence of feeding vegetarian primates a steady diet of such delicacies as sausages and beer.

It has been argued that as Europeans found themselves increasingly settled in large, sedentary villages and towns, and ultimately turned to urban fodder for the Industrial Revolution, they found themselves increasingly removed from "nature" and developed a kind of addictive craving for the presence of living wild animals. There may be something to that still. The average North American zoo visitor today is a middle-class, college-educated woman in her twenties, accompanying a child, attending a zoo in the spring or summer. Zoo surveys show that people want to see some active animals, they love to see baby animals, and if they could, they'd be happy to touch something. Petting zoos are enormously popular.

By the nineteenth century, Europeans had become positively manic for travelling troupes of freaks, monkeys, leopards, giants, jugglers, and other such wonders. But something sinister was at work. The extravagance at Regents Park was also about displaying the wealth and majesty of empire. That's the other thing about grand bestiaries. They serve an ancient, obsessive desire of warlords, magnates, and kings to acquire, bequeath, trade, and display exotic, strange, and magnificent wild animals.

From Kublai Khan to the Mughal emperor Akbar the Great, and from the thirteenth-century Holy Roman emperor Frederick II to American press baron William Randolph Hearst, bigshots of one stripe or another have always employed vast animal collections in dispensing and procuring loyalties and allegiances, favours, and alliances. Among the bears, monkeys, and other animals Charlemagne kept at his various residences was an elephant given to him by the Caliph of Baghdad. Phillip VI of France, Louis IX, and England's Henry II all kept extensive animal collections, trading specimens as tokens of esteem and fealty. To cement their newly established diplomatic ties in the sixteenth century, Russia's Ivan IV presented England's Queen Mary with "a large and faire Jerfawcon," and Queen Mary returned the compliment with a pair of lions. That tradition persists. In return for his 1972 gift of two muskox to the People's Republic of China, U.S. president Richard Nixon received a couple of pandas. Around the same time, Canada gave Saudi's King Khalid a present of gyrfalcons. The first collections at the Jurong Bird Park, when it opened in 1971, comprised mainly gifts from ambassadors.

In the ostentatious display of animals what's often at work is something deeply rooted in power, conquest, prestige, and domination. It can be as benign as the corporate sponsorship by Chemical Industries (Far East Ltd.) of the Malayan tiger viewing shelter at the Singapore Zoo (which sells such corporate sponsorships for 4000 Singapore dollars a year) or as savage as the excesses of the Roman emperors, who never flinched from the vices of debauchery. The Romans took the fetish to its most barbaric extremes. A single festival could involve the torture, maiming, and massacre of thousands of animals—bears, lions, giraffes, crocodiles, elephants, and bulls. If a few human beings were thrown into the bloodbath, all the more grisly and better the spectacle.

But captive humans would also sometimes suffice. In the sixteenth century, to impress the menagerie-fancier Pope Leo X, a Catholic cardinal kept a collection of humans. Among them were Tartars,

Africans, Indians, and Moors. The bestiary of Aztec emperor Montezuma, not long before, had included an array of dwarfs and "human monsters" among his animals at Tenochtitlán. Europe's medieval travelling menageries were often complemented by displays of humans, and as recently as the nineteenth century, English people were regularly coaxed out of a few pence to see collections of Laplanders, African bushmen, Ojibways, and Inuit. Such travelling shows routinely included even more exotic human specimens: bearded women, "boneless" children, giants, and even less fortunate people billed merely as "humanoids." The carnival, the freak show, the zoo—the lines were often blurry.

Even Carl Hagenbeck, the German bestiarist considered the founding father of the modern open-plan zoo—his naturalistic 1907 Hamburg Tierpark was the first to experiment with the concepts so elaborately perfected at the Singapore Zoo and Night Safari—was not above such vulgarities. The great innovator who pioneered methods of training animals through conditioning rather than beatings and privation once presented the German public with an impressive collection of human specimens from 36 ethnicities, including Sudanese, Patagonians, and Eskimos.

Around the same time, the New York Zoological Society's William Hornaday, who so passionately decried the extinction of species, oversaw a display at the Bronx Zoo that included hundreds of humans. Among them were Kwagewlths from the British Columbia coast, Zulus and pygmies from Africa, and Igorots from the Philippines. A popular exhibit at the time was Ota Benga, a pygmy from the Congo, who shared a cage with an orangutan named Dohong. Another was Geronimo, the famous Apache war leader who bravely fought invading American forces in the late nineteenth century. Geronimo died shortly afterward in 1909, at Fort Sill, Oklahoma, still technically a prisoner of war. Benga, after enduring a brief life as a travelling exhibit, hanged himself in Lynchburg, Virginia, in 1916.

After all these years, just when we thought we had clearly demarcated the lines between freak show and menagerie, zoo and wilderness, and human and animal, order is collapsing again. It is collapsing in ways no one foresaw, with the possible exception of such early science-fiction writers as H.G. Wells.

The argument for building an ark by using zoos, parks, aquariums, and aviaries, or even by relying on the contents of test tubes in the deep-freeze units of the British Natural History Museum in London, rests on some questionable assumptions.

One assumption is that human populations will one day stabilize at some tolerable abundance that will leave room for all those other forms of life. As it turns out, this is not unrealistic. Human population growth is occurring at different rates around the world. In Europe and North America, the population would be shrinking but for immigration, and worldwide, growth rates are starting to level off. But that does not mean that any time soon we'll all be living within our ecological means on a planet where the forests have grown back and the global climate has resumed some sort of stable pattern.

Michael Soulé, founder of the Society for Conservation Biology and chair of the Environmental Studies Board at the University of California at Santa Cruz, concludes that we have already entered a "demographic winter" that will last at least two centuries. After that, if human populations have not already imploded from disease or famine, or from wars fought over the world's dwindling natural resources, there will be a period of ecological restructuring that will last who knows how long. And maybe, after all that's happened, the non-human world can be gradually repopulated.

It is also unclear whether the drastic assumptions about the inevitability of species extinctions are justified, and whether radical intervention and "triage" are applicable only in such places of the world as the humid tropics.

When it comes from the zookeeping fraternity, the "ark" justification for zoos often has the faint whiff of casuistry about it.

Growing up on Canada's west coast helped me become inured to arguments zookeepers sometimes make. Vancouver's Stanley Park Zoo insisted that keeping whales in big swimming pools and training them to do endearing and humanlike things was all for the greater good. Every time the Stanley Park Zoo captured another killer whale for the aquarium trade, and every time one of its own killer whales died, and every time one of its killer whales' calves died, it was always the same. It was for the greater edification of the public, to more effectively inculcate a concern for the great whales of the seas. By the 1990s, British Columbia's distinct southern subspecies of killer whales was endangered, and its numbers continued to fall. The proximate causes of the killer whales' peril were the dwindling salmon stocks upon which the whales fed and the contaminants that had made the whales' bodies comparable to toxic waste. But the subspecies had not been helped by having had its numbers thinned in the first place by the lurid business of catching whales to supply the Stanley Park Zoo and other whale-show attractions around the world. During the 1960s and early 1970s, 62 killer whales were captured for the aquarium trade. A dozen died during their capture. Of those that survived, many died after only a short time in their tanks.

The Stanley Park Zoo also housed polar bears. The last one was Tuk, who died in 1997 at the age of 36. I have a vivid memory of being 12, perhaps 13, and watching a polar bear robotically pace in a concrete tank sculpted to look like a piece of iceberg. The bear simply paced back and forth—it was an automaton, not an animal—and I remember being possessed of an overpowering desire for it to be killed and put out of its misery.

The bear's condition, known as *stereotypy,* is a common ailment among zoo specimens. It's a kind of madness that is found among almost all mammal species, in most zoos, and is only one of several pathologies that come with captivity. Cannibalism, self-mutilation, hypersexuality, eating disorders, and a whole range of pathetic behaviours afflict as many as one-third of all the mammals in North American zoos. This was so even in "progressive" institutions, such as the famous San Diego Zoo.

For all the captive-breeding efforts of the Singapore Zoo, the Jurong Bird Park, and most other "progressive" zoos around the world, the work is mainly about keeping zoo populations reproductively healthy and transferring animals among and between zoos to further that purpose. In the Canary Islands, at Tenerife, several captive-breeding programs are underway at the opulent Loro Parque, home to thousands of specimens from hundreds of species. You can stroll through a glass tunnel while sharks and manta rays swim about you, see chimpanzees in their painstakingly reconstructed African micro-habitat, and view the Bengal tigers on their island in an artificial lake. In the oil-rich sheikhdom of Qatar, which remained outside the protocols of the Convention on the International Trade in Endangered Species (CITES) until 2001, a private menagerie, owned by Sheikh Saoud bin Mohammed bin Ali al Thani, contains thousands of animals. Many of them are from endangered species, and al Thani employs entire teams of veterinary scientists and biologists in his various captive-breeding programs.

At Tenerife, a small community of endangered gorillas provides an important genetic resource for the European Breeding Program's plans for the gorilla's "future insertion into the wild." And the Singapore Zoo did embark on a minor effort to reintroduce mouse deer and otters to Singapore's postage stamp–sized protected areas, though it was largely a public relations exercise. The program initially included a plan to reintroduce civet cats and leopard cats, but that idea was dropped on the more sensible counsel that it would probably result in an extirpation of the remnant populations of the cats' prey species.

It is only rarely, however, that captive-breeding programs are about the recovery of freely sustained, self-supporting, and viable populations outside of zoos. And rarely have those efforts succeeded, anyway. Attempts to resuscitate near-extinct animals go back to the sixteenth century, when the Polish nobility established huge wildlife reserves in hopes of saving the aurochs, the wild ancestor of most modern cattle. It didn't work. Only a handful of near-dead species can be said to have revived, even from local extirpation, even partly because of purposeful reintroduction.

One clear success story is the Przewalski's horse, a rugged little equine native to Mongolia that is likely the evolutionary ancestor of modern horses. There were none left in the wild in the 1950s, but following an ambitious reintroduction program based on small, domesticated herds, today about 1000 range the Mongolian steppes. The case of Père David's deer is difficult to situate in the discussion. Named after the French missionary Père David Armand, this species of Chinese deer was extinct in the wild for several centuries, and by the nineteenth century there was one remaining herd confined to the Chinese emperor's hunting park. Père David Armand obtained permission to ship some off to Europe, which proved fortuitous, since there were none left in China by the early 1900s. The 18 surviving European specimens were gathered at Woburn Abbey by the Duke of Bedford, and from this small herd, a population of 600 grew. But all of these deer are in various zoos around the world.

Most of these valiant efforts remain, at best, ongoing experiments. Such experiments are underway with the Arabian oryx, the peregrine falcon, the golden lion tamarin, the California condor, the Alpine ibex, the Aleutian goose, and the European bison, also known as the wisent, which survives in small herds in a forest reserve in Poland. For most endangered animals, zoos are hard to think about as stations on some trail back to Eden. Zoos are where they go extinct.

The quagga, a zebra-like creature known only from a handful of photographs taken of a single animal—a captive of the London Zoo bought from an animal dealer in 1851—ended its existence as a species in the form of another single animal, which died in a Dutch zoo in the 1880s. Pink-headed ducks from India appear to have died as a species with a single pair in the private collection of a member of the London Zoo's governing council, during the Second World War. The ducks were last observed in the wild in 1936, in India. Expeditions to remote northern regions of Myanmar and Tibet have failed to find any.

Even North America's great wilderness parks have a lousy track record in conserving wildlife. Environmentalists learned far too late (and many still haven't figured out) that small and isolated "representative" samples of wilderness, scattered across vast landscapes where untrammelled resource extraction otherwise proceeds apace, do far more harm than good. They provide a release-valve service to economies that require their firing chambers to be continually stoked with trees for lumber, mountains for minerals, and land for real estate development. Concessions in the form of parks, wilderness reserves, and marine protected areas have too often performed precisely that service, and it's one of the reasons why Montana conservationist and author Richard Manning astutely observes that a new park is never a victory—it's really an admission of defeat.

Biologist William Newmark helped to make Manning's case in a seminal 1987 study of the great parks and wilderness areas of the North American west. By exhaustive analysis of several decades' worth of wildlife records in the parks, Newmark showed that the parks had suffered a slow death of "mammalian faunal collapse." The parks were too small and the mammal populations they supported were too fragile to remain viable over the long term. More importantly, the parks were too isolated from one another to allow their mammals to maintain viable populations. Endangered animals won't be saved by confining them to fragments of habitat types that parks are intended to represent, Newmark found. What's needed are bigger parks, more parks, and the careful restraint of human activity within carefully managed wildlife corridors connecting the parks.

On the side favouring zoos as arks, as a temporary solution, is journalist Vicki Croke, an animal enthusiast but no naive advocate. In her 1997 book, *The Modern Ark: The Story of Zoos,* Croke makes a good case that the loss of genetic diversity that has befallen so many vertebrate species, especially birds, is irreversible, and it's time we faced it. There is no real wilderness left, she says, and what remains will soon be gone: "The wild world is becoming a series of mega-zoos." She is positively enthusiastic about cryogenic methodologies and test-tube solutions to the problem of extinction. Close the zoos, Croke says, but reopen them all the next day, with more resources and a different mandate. Make them arks. Just keep selling tickets.

On the other side of the debate is a formidable intellectual who, like Croke, does not come from the disciplines of zoology, the veterinary sciences, or the ecological sciences. David Malamud is an associate professor of English at Georgia State University. His 1998 book *Reading Zoos: Representations of Animals in Captivity* looks closely at the ways we have understood zoos and "constructed" them in our imaginations. Malamud casts a cold eye on what zoos mean, in broad cultural terms. He's against zoos and he draws little distinction between old-style collections of pathetic animals in smelly cages and the more modern game parks and open-plan institutions of the kind so well represented in Singapore. "I consider differences among zoos cosmetic," Malamud writes, "or otherwise insignificant in terms of mitigating their collective implication in a culturally retrograde enterprise."

His is not a recent viewpoint. In 1913, the novelist Thomas Hardy wrote a letter to *The Times* expressing his disgust with the very idea of zoos, calling them "useless inflictions" where animals were made to perform silly tricks or otherwise "drag out life in a wired cell." John Galsworthy, Hardy's contemporary and fellow novelist, saw in zoos the most vulgar and hypocritical state of humanity. They were "a horrible barbarity," he wrote.

David Hancocks, director of the Open Range Zoo at Werribee, Australia, has a different vision of what the zoo might be. In his 2001 work *A Different Nature: The Paradoxical World of Zoos and Their Uncertain Future,* Hancocks sets out the complexities and contradictions associated with our unquestionable love of other life forms. Some people express that desire by fighting to protect forests that support bears. Others express that love by hunting bears. Go figure. As for zoos as arks, Hancocks is unambiguous. The idea is "ludicrous," he says. "Zoos are not, and for many reasons cannot be, sanctuaries for saving the world's wildlife: they deal with too few species and too little space for it.... We cannot save the world's endangered wildlife through the few successful breeding programs in zoos, just as one cannot save a language by simply holding onto a rare document."

He makes a strong argument. The United Nations' Global Diversity Assessment counts roughly 5400 known animal species that face some threat of extinction. If all the zoos in the world gave over fully half their resources to the work of maintaining viable breeding populations of endangered species, Hancocks reckons they might hold on to perhaps 800 species.

Hancocks proposes that we "uninvent zoos as we know them and ... create a new type of institution, one that praises wild things, that engenders respect for all animals, and that interprets a holistic view of nature." Further, zoos must become diplomatic missions for the animal world, in the world of urbanized *Homo sapiens.* They should nurture the strong public desire to conserve wildlife habitat and incorporate conservation strategies as their defining organizational principles. These kinds of zoos, and no other, should be the only zoos "allowed by law."

Such zoos as Hancocks proposes already exist. The New York zoos administered by the Wildlife Conservation Society (WCS), including the Central Park, Bronx, and Queens zoos, as well as several other local institutions, are proof that captive animals can be put to more useful, honest, and humane purposes. More than four million visitors attend these zoos every year, and the simple zoo-going experience is vastly augmented by the institutions' far-reaching education programs. The WCS zoos also serve the diplomatic-mission purpose Hancocks proposes, by linking New Yorkers to projects the organization is undertaking in 53 countries around the world. Those projects are geared mainly toward securing necessary habitat for endangered species, "from butterflies to tigers."

WCS scientists are training conservation officers in East Africa. They have rediscovered a wild pig, thought to be extinct a hundred years ago, in Southeast Asia. They're working with government bureaucrats in Iran to conserve the last tracts of habitat frequented by the endangered Asiatic cheetah. They're mapping Amur tiger habitat on the Russian–Chinese border, developing conservation initiatives for the critically endangered crested iguana in Fiji, and using radio tags to track great white sharks off the coast of South Africa with satellites. Even the Singapore Zoo is starting to do this kind of work, establishing a Sumatran rhino research station in Sabah, on the island of Borneo.

This is not the work of building arks. It's about something far more hopeful.

A bird

Waiting for the Macaws

While the bird was asleep they carried it up to their bothy,
kept it alive for three days, and then killed it with a stick,
for there had been heavy gales not long before
and it might have been a witch.

—An account of the last great auk in the Outer Hebrides,
found by four men at Stack-An-Armine,
an islet off St. Kilda, in 1821

There are people whose love of birds will cause them to spend months tramping across mucky Scottish heaths hoping to see a red kite, or years thrashing around the Volga River delta looking for curly pelicans. They save their pennies for a once-in-a-lifetime trip to South Africa's Blyde River Canyon just for the slim chance of seeing a black-rumped buttonquail. Birdwatchers think nothing of giving over their entire lives to such things. They endure great hardship. They persevere. They stay awake.

This is the kind of thing I kept telling myself, sitting in a patch of dry forest swamp at a makeshift observation post more or less in the middle of the 1500-hectare Curú National Wildlife Refuge, on Costa Rica's Pacific coast. I'd travelled 5800 kilometres for a glimpse of a scarlet macaw. I was in the right place, at the right time. But the iguanas were going lazily about their business, climbing up and down the coconut trees like giant languid squirrels, and the air was growing warmer and drowsier with the scent of hibiscus. Huge land crabs slowly clattered to and fro across a carpet of dead palm leaves, and strange blue butterflies fluttered in and out of the forest canopy. I could barely keep my eyes open. This wasn't exactly hardship.

Every few minutes a coconut fell to the ground with a thud, and I'd be alert again to the fact that if I was ever going to catch a glimpse of a scarlet macaw in its own element ever in my life, then it would happen right here. I'd been directed to the spot by Greg Matuzak, a lanky, 34-year-old Californian biologist who oversees bird research projects at Curú. Bernadette Bezy, a marine ecologist who had also come to Curú from California, had lent me a pair of binoculars. Matuzak had even graciously given me two 250-gram packages of Kiki Girasol sunflower seeds to empty into two boxes hanging from poles in the clearing. The macaws come to the boxes between three and four o'clock every afternoon, Matuzak had said. Sunflower seeds are like macaw candy, and macaws are smart, so all you have to do is sit quietly in the grove of palm trees in that patch of dry swamp, about 25 metres from the seed boxes. There are camp chairs there and everything. Nothing to it.

But the reason I'd come to Curú was not just for a glimpse of an especially beautiful kind of parrot. It was because events unfolding at Curú were showing that the fate of the world's vanishing birds was not necessarily extinction, aviaries, or the limited confines of a few parks—the trends in the extinction of the world's "wild" things do not have to lead in the same dismal direction.

Curú is a paradise of tropical forest, orchard, and pasture at the southern tip of the Nicoya Peninsula, at the end of a bumpy road a few kilometres south of the town of Paquera. Forty years earlier, there had been no scarlet macaws left at Curú. The birds the locals call *lapa roja* and scientists call *Ara macao* had vanished from pretty much the entire Nicoya Peninsula—and indeed, by then, from almost all of Costa Rica. Once common throughout the Caribbean lowlands on the other side of the country, the macaws had disappeared there, too, except for a few near the Nicaraguan border.

Throughout the Nicoya they had flocked in communal roosts of as many as four dozen birds, making their nests in the cavities of gallinazo trees or in the deep hollows of ancient ceiba trees, revered by the Maya as providing a mysterious conduit between the sky world and the underworld. Most of that old forest was gone, and the scarlet macaws with it. Over the years, the birds had been killed for food, shot out of the trees by farmers who considered them pests, or captured for the pet trade. But now, after all these years, macaws lived at Curú again—not many, but some. It was a heartening departure from the clear trend among the world's birds.

The best information the International Union for the Conservation of Nature (IUCN) has about any kind of animal is about birds. Roughly 10,000 bird species exist in the world; about 1 in 8 is threatened with extinction. Of those, 250 species are considered "critically endangered," a status set aside for species with a 50-50 chance of vanishing within five years.

Among those hard cases is the crested shelduck, a favourite subject of early Japanese watercolours. The bird was once found over a wide stretch of North Pacific coastline, from Vladivostok to South Korea. If unconfirmed reports from peasant farmers on the Chinese mainland coast are anything to go by, perhaps 50 pairs are left, but no confirmed sightings have been recorded since 1964. In similar straits is the northern bald ibis, the bird the ancient Egyptians considered holy. This splendid bird was once widespread across Europe, and its rookeries were known from Syria to Switzerland. It has been reduced to roughly 220 birds, in 3 small breeding colonies in Morocco. The po`ouli, the little Hawaiian honeycreeper I had wanted to write about, was even more unlucky—it had become extinct just months before I came to Curú to look for macaws and talk to Greg Matuzak about parrots.

Of all the great bird families, the parrots are distinguished by the highest proportion of species—close to one-third—threatened with extinction. The order Psittaciformes encompasses roughly 340 existing species of cockatoos, conures, lories, lorikeets, macaws, parakeets, and parrots. The terms get a bit confusing because the word *parrot* is usually used to describe all these species but sometimes it's used in a way that puts "parrots" and macaws in the Psittacidae family and shuttles the cockatoo species, the lories, and lorikeets into the Cacatuidae and Loriidae families. For our purposes, we'll be using the word *parrot* to mean all the Psittaciformes.

The smallest is the buff-faced pygmy-parrot of New Guinea, which looks exactly how you would imagine a green parrot to look, except it's about the size of a hummingbird. It rarely exceeds eight centimetres from its beak to its tail. The largest, by weight, is the solitary, flightless, four-kilogram kakapo of New Zealand. It looks a bit like a big green owl, growls like a dog, and has been known to hike several kilometres on its nightly foraging rounds. Through the ages it developed a predator-avoidance strategy of standing perfectly still in the hopes of going unnoticed. That habit didn't do it much good when the Maori and their dogs arrived in New Zealand about a thousand years ago and was of no better use after Europeans arrived with their cats. Fewer than 100 kakapos are known to exist today.

Of all the different sorts of parrots, the macaws have suffered particularly badly. As many as eight macaw species were already extinct by the beginning of the twentieth century. As recently as the late 1980s, 3 of the world's 17 remaining macaw species were regulated by the

Convention on the International Trade in Endangered Species (CITES); 15 years later, all 17 are on the convention's lists.

Macaws are the big South and Central American parrots, the ones you saw sitting on the shoulders of pirates in all those old movies when you were a kid. They're long-lived birds—they've been known to live at least 60 years in captivity. Long before Europeans developed a fascination for macaws, South and Central American peoples commonly kept the birds as pets. Like all parrots, macaws are highly social animals, and they're famous for their oddly humanlike characteristics and their astonishing capacity for mimicry. Or maybe macaws have noticed that people are highly social animals with oddly macawlike characteristics, such as an astonishing capacity for mimicry. Either way, people and parrots have long held an affinity for each other. It is a sad paradox that this shared affection is one of the main reasons so many of their species are in danger of disappearing.

One of the 17 macaw species on the CITES list, the glaucous macaw, is still officially listed by the IUCN as endangered even though it hasn't been seen since the 1960s. It is—or was—an exquisite, grey-headed, turquoise-feathered bird. Its numbers fell through the twentieth century in tandem with the rise of the rare-bird trade and the vanishing of yatay palm groves on the frontiers of Brazil, Paraguay, and Argentina, its only known habitat. British bird conservationist Tony Pittman undertook two extensive expeditions to the region during the 1990s and failed to turn up even a rumour of one persisting anywhere in its former range. Another species, Spix's macaw, became extinct in its home habitat only nine months into the twenty-first century. This leaves 15 macaw species with at least some of their members persisting, however tenuously, outside of aviaries and private collections.

The "blue" macaws have fared the worst. Like the glaucous and Spix's macaws, the hyacinth macaw is a blue macaw, although it's actually a striking cobalt. It was once common throughout much of

Brazil and from Bolivia to Paraguay. During the 1980s, about 10,000 hyacinth macaws were bought and sold in the big-money bird markets of the world. The twentieth century ended with perhaps 2500 remaining outside of cages, in Brazil. Another blue macaw is the Lear's macaw, which was first described to science by Prince Charles Lucien Bonaparte, Napoleon's nephew, in 1856. The prince named the bird after Edward Lear, who was one of the most accomplished bird artists of the nineteenth century, a fact usually overshadowed by the nonsense verse that made him famous.

Lear's macaws had been making their way into the hands of European bird collectors since the early nineteenth century, but they were one of ornithology's enduring mysteries. The source of the birds was unknown to science until the late 1970s, when Helmut Sick, a young German biologist who first came to Brazil on assignment with the Berlin Museum, solved the puzzle. In 1939, Sick had been sent on an expedition to seek out the red-billed curassow in the wilder corners of the Brazilian state of Espiritu Santo. The Second World War broke out, and Sick's planned three-month sojourn stretched into six years. He spent the first three years in hiding and the last three in prison, as an enemy alien. After his release, he decided to stay put and settled in with the Museu Nacional in Rio de Janeiro. Then, in 1954, Sick went looking for Brazil's mysterious Lear's macaw. His search lasted 24 years.

On December 29, 1978, a triumphant Helmut Sick came upon a population of a few hundred Lear's macaws in the remote sandstone canyons of the Raso da Catarina area of Bahia, in northeastern Brazil. While I was sitting in that patch of dry swamp at Curú, undergoing privations no greater than trying to keep awake, permanent guards kept watch over the only two remaining nesting sites of Lear's macaws in the world, in Raso da Catarina. The guards were employed by the Brazilian conservation foundations Biodiversitas and BioBrasil. Only 240 Lear's macaws remained.

The most infuriating blue macaw story involves Spix's macaw, which until October 2000 made its home in the unforgiving Caatinga region of northern Brazil, a harsh and dry Texas-sized landscape of sparse forest, cactus, and thornbushes. Spix's macaws were never very plentiful, being confined within the Caatinga to the woods along a few rivers that are dry for most of the year. The last female in the Caatinga was captured by poachers on Christmas Eve, 1987. That left a single male.

The species had been named after Johan Baptist Ritter von Spix, a Bavarian naturalist who travelled throughout Brazil in the early nineteenth century. Spix's companion, Carl Freidrick Philip von Martius, shot a macaw on the banks of the São Francisco River in 1819, adding what he thought was a hyacinth macaw to the specimen collection he and Spix were gathering. Their collection would grow to represent about 350 bird species before their four-year South American journey ended in 1820. It wasn't until 1832 that the beautiful blue parrot Martius shot on the São Francisco was discovered to be something else altogether, something exceedingly rare. Munich zoologist Johann Wagler compared the São Francisco macaw with specimens of hyacinth macaws and found that it was clearly smaller, and it had black skin on its face and a greyish head. It wasn't just a new species, either: it turned out to be the sole occupant of its very own genus. Spix's macaws quickly became a prime collector's item among wealthy nineteenth-century European bird collectors. But few aviarists ever got their hands on one.

Expeditions to Brazil in search of the Spix's macaw routinely returned without any birds. Collectors often had trouble finding anyone in Brazil who'd even heard about them. Between 1820 (when Spix and Martius returned to Germany) and the 1970s, ornithologists observed the birds in the wild only twice. It wasn't for want of trying, either.

By the middle of the twentieth century, little remained of the precarious ecological niche the birds inhabited—the gallery woodland of caraiba trees in Brazil's dry and ravaged Caatinga. Three of the birds were believed to exist there in the 1980s, but that was just a guess. And after Brazil banned the export of wildlife in the 1960s in an effort to stem the depletion of its rare and endangered species, and especially after CITES prohibited the trade in Spix's macaws in 1975, it became just as difficult to determine how many Spix's macaws existed in captivity.

This presented conservationists and the Brazilian government with a conundrum. To save the bird, effective breeding pairs would have to be found among the captive birds. And so without an arrangement of some kind with the few obsessive millionaires, eccentrics, and pathological acquisitors who held Spix's macaws in their collections, an effective species-recovery strategy would be impossible. The Brazilian government, along with conservation groups and CITES officials, were forced into a devil's bargain.

Worldwide, the illegal market in rare and exotic birds often does as much harm as habitat loss. It isn't uncommon for a bird that is desperately rare in the wild to be fairly common, however temporarily, in the pet trade. The blue-throated macaw had been reduced, by the 1990s, to perhaps 50 birds in its only known habitat, in a remote corner of Bolivia, but several hundred existed in private collections. By the 1990s, the great green macaw, also called Buffon's macaw, was similarly endangered, despite a former range that takes in a half-dozen countries, but hundreds survived in private collections.

The rarer the bird, the higher its market value. That's the logic driving both the legal market and the black market in rare birds. The legal market is based on the trade of endangered birds in captivity prior to the CITES restrictions, as well as the trade of their offspring. The black market is supplied by gangsters who raid nests in the forests of the world and "launder" the birds with false documents. Supply and demand grind extinction's treadmill. A diminishing supply boosts the demand for rare parrots, which makes them rarer still, which raises their price, and on it goes.

By the 1980s, a single Spix's macaw fetched $30,000 U.S. on the black market. In the early 1990s, a Brazilian dealer sold a single Lear's macaw for $13,000 and a new car. A decade after that, a single hyacinth macaw could readily fetch $10,000 on the open market. While I was writing this chapter, it took me less than 15 seconds to find a pair of hyacinth macaws advertised on the internet for $28,000.

The underground trade in wild birds is a vile business that's often as lucrative as high-stakes international drug trafficking and gun running. It is also quite separate from the conventional trade in, say, budgerigars—endangered-bird traffickers and collectors are of a different class altogether from the millions of largely harmless people worldwide who keep pet birds. This isn't to say the lines are always clear. Many conscientious bird enthusiasts unwittingly contribute to their subjects' peril by buying endangered birds that originated in the black market. Many rare-bird collectors profess transparently disingenuous concerns about the fate of their birds' free cousins. Still other collectors are quite obviously and sincerely concerned about conservation and species recovery.

It's never easy sorting these things out, and in the late 1980s the International Council for Bird Preservation had few rules to rely on in its dealings with bird collectors. The CITES specialists' group on parrots was contemplating a similarly blank page. In the midst of this uncertainty, the basis for the Spix's macaw recovery plan was established, and the devil's bargain was struck.

Informed speculation about the number of Spix's macaws in private hands and in zoo holdings put the worldwide population of captive Spix's macaws at perhaps two dozen. At the time, it was uncertain whether any persisted in the wild. Spix's macaws had been successfully bred in captivity perhaps twice, and outside the world of private zoos and aviaries there was little expertise in the all-important work of captive breeding. Brazilian government officials faced the cold truth that no recovery program for Spix's macaws could work without the willing participation of the bird collectors who held those that remained. So they offered an amnesty in exchange for the private owners' agreement that all buying and selling would come to an end and all birds would be made available for a carefully controlled breeding program, geared to the recovery of the birds in the wild. By 1989, the Brazilian government, with the cooperation of CITES and bird conservation groups, had cobbled together an ad hoc international committee to oversee the work. The committee first met that year at the CITES annual meeting, in Lausanne, Switzerland.

The committee's membership included some individuals with questionable backgrounds, such as Tony Silva, an American who had been selling parrots into the United States from Antonio de Dios's "parrot factory" in Quezon. At the time, it later turned out, Silva was engaged in a hugely profitable business smuggling hyacinth macaws and other endangered birds from Brazil to buyers in America (in 1996, Silva was convicted in a U.S. court for violating the U.S. Endangered Species Act and sentenced to a seven-year jail term). Also at the meeting in Lausanne was Antonio de Dios himself—a Filipino millionaire whose heavily guarded private bird-breeding facility in Quezon hosted several thousand parrots from almost half the world's species, including a half-dozen Spix's macaws. Wolfgang Kiessling, a man not without a conscience, was also in attendance, as the owner of two Spix's macaws. Kiessling kept them at his Disney-scale zoo, aquarium, and aviary at Loro Parque, on Tenerife—the Canary Islands "ark" described in the previous chapter. Lurking in the background at the Lausanne meeting were other figures, some of whom were associated with a macaw-smuggling enterprise in Paraguay.

The committee's work lasted a little better than a decade. During its term, it established the whereabouts of more than 60 captive Spix's macaws, but most of them turned out to be closely related and were consequently poor prospects for founding a new population. The committee's biggest boost came during its early days, with the July 9, 1990, discovery of one last Spix, a male, in the Brazilian Caatinga, in a patch of caraiba trees at Malencia Creek. Elaborate plans were laid out and enormous sums of money spent on plans to return some of the captive birds to join the last wild Spix. Kiessling personally donated more than $500,000 U.S. to the effort.

In the first of many setbacks, a female Spix that showed some promise of becoming a mate was released at Malencia Creek, only to die entangled in electrical wires on a power pole. Then, on October 5, 2000, that last male Spix, after having survived alone in the Caatinga for 14 years, disappeared. The poor farmers of the region undertook a massive search for him. They had rallied behind the bird through the 1990s, taking him on as a symbol of their own hardscrabble persistence in that parched countryside. Their search was in vain.

The recovery committee collapsed a year later, and no honest person shed a tear over its demise. It had degenerated into a private trading club among millionaire Spix fanciers. The one glimmer of hope was the effort Wolfgang Kiessling was putting into his own breeding program on Tenerife. Kiessling surrendered the ownership of his three macaws to the Brazilian government, and in return, the Brazilian government allowed him to hold the birds and cooperated fully with his breeding program. In 2005, a chick was born at Kiessling's facility on Tenerife. The Brazilian government possessed nine other macaws, but there wasn't much talk about a reintroduction program. Apart from two birds held by a Swiss collector, the world's remaining macaws—all 42 of them—had been acquired by billionaire Sheikh Saoud bin Mohammed bin Ali al Thani, the Qatari menagerie-owner we met briefly near the end of the last chapter.

As much as all this was enough to make you want to spit on your hands, raise the black flag, and start slitting throats, it was all the more reason to take heart from what was happening in Curú and, indeed, throughout Costa Rica.

At Curú, the very thing the conservationists trying to save the Spix's macaw had only dared to hope for was taking place—a kind of life-after-death story. Greg Matuzak and his volunteers were showing at Curú that not only can the world's vanishing birds escape extinction, but they are not inevitably destined to the end-of-the-road existence of zoos, aviaries, and parks, either. There is life after captivity.

The Curú experiment is still underway, but it has already shown that macaws can be re-established where they were once locally extinct, and even "hand-reared" macaws, under the right conditions, can make a go of it back out there where they belong. All the scarlet macaws at Curú had been born captives. Their parents had been confiscated from poachers by the Costa Rican government. The birds had been hand reared at a private facility at Alajuela, near San José, Costa Rica's capital city, and now they were free and starting to raise their own young. It was the first time that hand-reared macaws reintroduced to the wild had successfully borne offspring.

"You'll probably see one. I don't know why not," Matuzak said. We were strolling through Curú, on a dirt track following the edge of an old pasture that was slowly being taken back by the forest. We passed a rock-and-mortar shrine to St. Francis of Assisi, the Catholic heretic/saint who preached that birds have souls, and continued on toward a muddy creek in one of the pastures where the cattle come to drink. Only the other day, some scarlet macaws were bullying a bunch of black vultures there, Matuzak said. They might be there now, you never know.

When we got there, about a dozen vultures were hanging around, and a family of white-tailed deer moved through the long grass. So we decided to wait around for a while, and Matuzak went through the complicated story behind the return of scarlet macaws to Curú.

In January 1999, 13 scarlet macaws born at Alajuela were brought to Curú. They spent two months in huge cages near the seed-box observation post where I'd been waiting for them, and for another six months they were slowly weaned from a diet of fruit, beans, rice, and dog food. Apart from their minor daily treats of sunflower seeds, the

13 birds quickly became fully dependent upon the fruits of the forest. Before long, they were happily foraging on the flowers, seeds, bark, and fruit of 25 tree species, such as wild plum, beach almond, jocote, ojoche, coconut, and royal palm.

The Curú project, sponsored by the bird conservation group Amigos de las Aves, was undertaken in conjunction with two other scarlet macaw release efforts. One was at Golfito, a valley adjacent to the 15,000-hectare Piedras Blancas National Park on Costa Rica's southwest coast. The other was undertaken at Tambota, in Peru, at a remote site adjacent to two parks comprising almost 800,000 hectares. A major point of the whole exercise was to determine what works and what doesn't in bird reintroduction programs of this kind. Comparatively, the Curú experiment was more successful than those at Golfito and Tambota. At Golfito, 34 birds were released, 2 of which were killed almost immediately by an ocelot; a third was eaten by an eagle. Still, four years later, half the original Golfito birds were still believed to be alive. At Tambota, 11 of 20 were alive four years later. At Curú, 9 birds survived the initial release of 13. Then, in 2004, 2 chicks were hatched, bringing the population up to 11.

The birds at Curú had names, thanks to Fiona Dear, an English biologist whose work with Matuzak had been sponsored by such outfits as British Northern Parrots, the Paradise Wildlife Park, and a charity fund maintained by the Cadbury's chocolate company.

Emilio and Talula were the lucky ones who'd produced the two fledglings, Hans and Rita. There were Eva and Renaldo, and another pair, Cariño and Jemima. *Cariño* means "caring" in Spanish, and he got the name because he's something of a sweetheart, especially to Jemima. Ringo was so named because he had a ring, or band, on his right leg rather than on his left, as the others did. He tended to be at the bottom of the social ladder, the one that waited at the seed boxes until everybody else had eaten their fill. Rico was the most beautiful of the lot, they say. The last one was unnamed. He was a solitary character, though he used to hang around with Cariño and Jemima. He kept his own counsel and had shown up only a few times at the seed boxes.

The main point of the daily ration of sunflower seeds in those boxes was to allow biologists to monitor the macaws' health and their seasonal movements, to find out whether there were any fledglings around, to study relationships between the birds, and that sort of thing. It wasn't about feeding the birds. They were finding food all by themselves.

Another particularly heartening outcome of the Curú experiment was that the macaws were foraging outside the refuge. This opened up possibilities that you couldn't contemplate in Singapore, or anywhere else in Southeast Asia where habitat loss was the price exacted by the stoking of economic engines. In North America, William Newmark's studies had exposed the great parks and wilderness areas of the continent's western half as places within fragmented landscapes where animals go to die. At Curú, and in the landscape around it, a different kind of story was taking shape.

Curú was Costa Rica's first private wildlife refuge. After watching the forests disappear from much of Costa Rica in the 1960s, Federico and Doña Julieta Schutt, prosperous ranchers, had decided they didn't want to see the ruin of their beloved countryside. They started with an 84-hectare refuge and later won official "protected forest" status for all but a small portion of the remaining 1400 hectares of their former hacienda. The result is formally known as the Refugio Nacional de Vida Silvestre Curú. The refuge where these experiments are now underway is surrounded by a vast no-hunting zone of ranch country and broken forest—territory the macaws were venturing into.

The Curú experiments opened up the very real possibility that with a large enough population of macaws with their epicentres in refuges such as Curú, the birds might eventually repopulate much of their former range. Costa Ricans showed broad support for such ambitious projects; in fact, they had come to think of the scarlet macaw as a symbol of the country's unusual efforts to maintain the diversity of living things that the rest of the world was losing.

What allowed Costa Ricans to envision such possibilities was the fact that theirs is a relatively comfortable country that has sorted out the deep contradictions that ended up disfiguring so much of the tropical world—the very contradictions that result in mass extinction. Costa Rica is poor by conventional economic measurements, but it's also an island of peace and stability in a region that remains crippled from the rebellions, military coups, revolutions, and counter-revolutions that raged throughout much of the twentieth century. Hundreds of thousands of innocents died in El Salvador, Guatemala, Nicaragua, Panama, and Honduras, but Costa Rica avoided the bloodshed. It wasn't just through good luck, although that had a lot to do with it. It was also because during the 1950s, Costa Ricans forged a democratic consensus that was broad enough to enjoy the support of both the Catholic Church and the Costa Rican Communist Party. The consensus inhibited the development of Marxist insurgencies, but perhaps most importantly, a constitutional prohibition on maintaining a standing army immunized Costa Rica from the U.S. government's usual method of bullying, corrupting, and brutalizing Central American societies.

By the 1980s, Costa Rica had further distinguished itself by an earnest commitment to protect its ecological heritage. The country had lost most of its aboriginal peoples centuries before, mainly to disease. Those cultures never recovered. It had also lost more than half its forests to industrial logging, which gathered steam during the 1950s. Today, Costa Rica's national park system covers only 11 percent of the country, but roughly 25 percent of its land mass is at least nominally protected from high-impact development—a far higher threshold than most countries can claim. While only two of Costa Rica's ecological reserves are "absolute" reserves (Cabo Blanco, not far from Curú, is one of them), the other protected areas include vast contributions from private owners, like the Schutt family at Curú. Deforestation continues, but its pace is slowing. Costa Rica's famous cloud forests and other natural attractions draw millions of visitors to the country every year, bolstering the national economy: ecotourism ranks lower only than bananas as a contributor to the country's gross national product. The country is consciously making its way down the "sustainability first" road described by those UNEP scientists at the beginning of this book.

Costa Rica also started out with the distinct advantage of an immense ecological legacy of rainforest, dry forest, mountain, mangrove, swamp, savannah, marsh, and plain. The country is just over a third the size of England, but it hosts roughly 850 bird species—more than remain in all of Europe, and more than in all of North America. Even in the middle of the night, driving down the dusty and rutted roads of the Nicoya Peninsula, I was routinely shocked by the sudden darting of paraques into the headlight beams. I'd made a habit of taking an afternoon cappuccino outside the Iguana Café in the hippie-surfer town of Montezuma, not far from Curú, and during each visit I was harassed by flocks of half-metre-long, bright blue birds with long tails and ridiculous curved-feather crests jutting from their foreheads. They were white-throated magpie jays. The 511-page *Guide to the Birds of Costa Rica* describes their habits this way: "Travels in noisy, straggling flocks of 5–10 … loudly mobs Spectacled Owls and other predators, including man." I'll say.

Costa Rica owes its enviable avifaunal diversity partly to the inherent richness of the neotropics and partly to a fortunate geological history. About 50 million years ago, an archipelago of volcanic islands—hotbeds of biological isolation and speciation—began to rise up out of the sea to the south of a peninsula that jutted out of North America to roughly where Nicaragua is today. The islands developed distinct local forms of swifts, parrots, kingfishers, hawks, and other birds that colonized from distant lands. By about five million years ago, the isthmus of Central America was beginning to take shape, allowing diffusion and exchange of species between the two continents. Jacamars, toucans, cotingas, and manakins moved north. Gnatcatchers, motmots, swallows, quails, finches, and others moved south. Pacific and Caribbean species flourished on the coasts, and mountain and lowland species began to take on characteristics unique to themselves. When the great high-country forests took root, they became, for some birds, barriers to migration as effective as great expanses of ocean between islands. For others, the forests were like islands surrounded by fathomless seas. All this isolation and confinement further obliged immigrant birds to settle down and go about the slow and steady business of evolving into their own distinct forms. Unlike almost everywhere else, in Costa Rica hummingbirds sing. Some tanagers, on the other hand, gave up singing altogether. And the thousand or so scarlet macaws north of Costa Rica ended up in their own subspecies. Their colours and habits are slightly different from those of the scarlet macaws to the south, which occur in small populations as far distant as Brazil and Peru; Costa Rican scarlet macaws form a kind of transitional population between the two subspecies.

Costa Rica's most famous bird, the resplendent quetzal, is rarely described without superlatives. The male is often called the world's most beautiful bird, with its distinctive iridescent blue-green tail feathers that routinely reach more than 60 centimetres in length. The bird derives its name from Quetzalcoatl, a deity known to all Mesoamerican cultures but revered most fervently by the ancient Mayans, whose gods were usually a nasty lot. Quetzalcoatl—the feathered serpent—was different. He was a champion of ordinary blokes and was associated with mercy, charity, and liberty. In the pre-Columbian period, the quetzal's long tail feathers were trade items of immense value, reserved mainly for adornment by the Mayan nobility, who also traded quetzal feathers to the royal families of neighbouring kingdoms as far north as the Valley of Mexico and as far south as the mountain palaces of the Inca in Peru. The quetzal is Guatemala's national bird; it appears on the Guatemalan flag, and it gives its name to the country's main unit of currency. But it is only in Costa Rica that this bird has survived in reasonably healthy numbers. It is one of the main attractions in the cloud forests at Monteverde, the most famous of Costa Rica's forest reserves.

But other Costa Rican birds are just as beautiful. The violet sabrewing, a mere hummingbird, is a shy creature, despite being a giant among hummingbirds. It commonly reaches 15 centimetres in length, and the male's lovely green feathers and black tail are rarely noticed on account of its brilliant, almost deep purple colour. Another bird, the long-tailed manakin, is not unlike the quetzal in its ostentation. It's a small black bird with a sky blue back and an olive green breast, and the male is distinguished by long tail feathers that double its usual 12-centimetre length. The purple gallinule looks like a common coot dressed up for carnival. The tanagers are outrageously coloured in lime green, yellow, pastel blue, black, and mottled brown. The scarlet macaw, meanwhile—shocking bright red, blue-green, and yellow—is a breathtaking sight amid the emerald canopy of a Costa Rican forest. Or so I'd been told.

When Matuzak wasn't keeping track of the macaws, he was working on a study of the yellow-naped parrots that nest on the Tortugas Islands, just offshore. The parrots forage within the refuge at Curú and hang around in the mangrove swamp, and they are a worry. They're an endangered species; Curú has only about 150 of them. Matuzak was clearly smitten. He'd ended up overseeing the scarlet macaw project after a spell as a field biologist there, but he'd begun his time in Central America researching melodious blackbirds, blue ground doves, and yellow warblers in Honduras. Matuzak graduated from the ecology department at the University of California at Davis, but he'd started out in economics, at San Diego State University. The story is more complicated than that, of course. These stories always are. For Matuzak, who grew up in Redondo Beach, the thing about birds and other living things began when he was a kid, camping at Yosemite National Park with his dad, and fishing, hiking, and skiing. Somewhere along the way, the birds got to him.

Birds are particularly reliable witnesses for the case that human beings belong to a daft and bumbling species that nonetheless harbours an abiding affection for other creatures. From the ibis of the ancient Egyptians to the raven and the eagle that fly within the cosmologies of the aboriginal nations along the coast where I live, birds have always summoned our deepest longings and have always managed to lift the human spirit somehow. They get to you.

The birds that got to me were crested mynahs. There were once thousands of those lovely and harmless little birds on Canada's west coast. On the streets where I grew up, in Vancouver, Burnaby, and New Westminster, the little black birds with white patches on their wings used to gather in little flocks of a half-dozen or so. They would skip along down crowded sidewalks, looking just like busy little old men with their arms clasped behind their backs. Chinese immigrants had brought the birds to Canada in the 1890s, as affectionate companions from home. The few birds that escaped their cages over the years eventually established flocks, and those were the mynahs that had fluttered and chirruped through my own childhood. Tamed ones could be taught to repeat words. Street birds tolerated the company of people and amused themselves by mimicking the sounds of the city—doors opening and closing, the air brakes on buses.

But, like Singapore, Vancouver was constantly changing and growing, inventing and reinventing itself, demolishing its old buildings and constructing ever taller ones so that you could barely see the mountains anymore. There was more pavement, fewer trees, and more nest-robbing European starlings. By the beginning of this century, only two crested mynahs remained in Vancouver, a nesting pair. In February 2003, one of them was hit by a car at Second Avenue and Columbia Street, near the Cambie Street Bridge. The last mynah kept a faithful vigil beside its dead mate, until it too was run over, two weeks later. And then there were none.

It's cold comfort to be told that crested mynahs were not "native" to Vancouver, or that from an "environmental" perspective their disappearance did not diminish the integrity of the species, which still flourishes in Asia. The loss of Vancouver's mynahs diminished the city all the same. When I saw a crested mynah in Singapore, casually hip-hopping down the sidewalk outside the Sri Sri Foot Reflexology establishment in Arab Street, just around the corner from the In The Name Of Allah The Most Gracious The Most Merciful Mohd Rajeen & Brothers Café, I was overwhelmed by melancholy and a childish happiness.

Fleeting encounters with birds are known to cause in people something the American writer Joseph Kastner has called "surprised enchantments." Sometimes these are just profoundly personal, mysterious, and delightful little experiences that one doesn't spend too much time pondering because they are, after all, imponderable. At other times they strike like a shock of insight. They come as a consequence of completely random events, and it is as though all the notes of some magnificent, ancient chord have been struck, thoroughly by chance, deep within one's subconscious.

Birds have always soared in the vast skies of humanity's collective unconscious. Birdwatching is commonly called the most popular hobby on earth; American birders claim that they are 40 million in number and spend $32 billion every year in their pursuits. And birds sparked the modern worldwide struggle for the conservation of wild things: the first international covenant for the protection of wildlife was an 1895 summit of European nations in Paris, and its great result was a multilateral accord for the conservation of birds. Every international treaty on environmental protection owes its origin to that first covenant. The "father of ecology" is often identified as John James Audubon, the gifted French American bird artist. The environmental reawakening that occurred in the 1960s started with Rachel Carson's *Silent Spring,* a shocking overview of the worldwide effects of pesticides, most notably their effects on birds.

The recoveries of near-extinct birds are among the greatest successes of the conservation movement. The whooping crane, a regal creature that lost its tall-grass prairie habitat with the advance of farming across the United States, was down to 20 individuals by the 1940s, but because of the vigilance and devotion of its human admirers, it numbered about 200 by the beginning of this century. The California condor—misnamed, since it once ranged throughout much of North America—was once as badly off as the whooping crane, with about 20 animals left. Written off by most conservationists in the 1960s, the condor won the hearts of a few committed eccentrics whose perseverance allowed the population of North America's largest flying bird to grow tenfold by the late 1990s.

At the same time, of all the creatures that have disappeared because of human activity, birds have fared especially badly. One-fifth of all the birds that existed 20,000 years ago are now extinct, mainly because of people. Still, of the roughly 250 bird species that have vanished since the time of Christopher Columbus, more than two-thirds were lost not so much because of wanton slaughter, but because critical habitat had been taken from them, or their eggs had been eaten by human-associated species. This is the fate that befell the dodo, on the island of Mauritius in the Indian Ocean. Perhaps the best-known extinct bird of them all, the dodo, a flightless, turkey-sized creature, was subjected to wanton slaughter by passing mariners, but it finally succumbed to a host of torments, not the least of which were the ravages of several introduced species, including pigs, dogs, goats, and cats. Importantly, the dodo was the first animal understood to have been rendered extinct from causes at least indirectly attributable to human activity.

The last dodos were from a small flock observed by a marooned Dutch sailor in 1662.

Even in the most cursory review of the sad story of animal extinctions in modern times, what becomes quickly obvious is the prominence of island-dwellers. The reasons are complicated, and explained eloquently in David Quammen's brilliant and exhaustive *The Song of the Dodo: Island Biogeography in an Age of Extinction.* But the main reason is that island species tend to exist *only* on a single island or a small archipelago; when they're gone, all of their kind are gone with them.

For all we might observe about humanity's negligence and wantonness, over-hunting was the culprit in only about a tenth of bird extinctions over the past 500 years, and even in those cases people often fought desperately to save them. But those people, and their stories, have been largely forgotten. One such story concerns the great auk.

The cause of the great auk's extinction is no mystery: it was hunted to death. Slaughtered for its flesh, for the oil that could be rendered from its fat, and later for its feathers and skins, Europe's original "penguins" once roosted on rocky islets throughout the North Atlantic. By the early nineteenth century, apart from a few stragglers in Scandinavian waters, no auks survived on European shores. The last credible European account—of a solitary bird—comes from the Scottish island of St. Kilda, in the Outer Hebrides. For generations, sea-fowling villagers had taken great care to leave enough eggs in the auks' nests to replenish the great flocks, but by the nineteenth century the birds had become so rare that individual stragglers were regarded as possible omens. In about 1820, one had been captured and kept for four days, but it had escaped. The last auk at St. Kilda was found by four local men, on a skerry known as Stack-An-Armine, in 1840. It was also kept, for three days, but owing to unusually heavy winds and gales that accompanied its discovery, one of the men killed the bird with a stick—he thought "it might have been a witch."

The great auk's last great breeding colony was on Newfoundland's Funk Island. During the latter half of the eighteenth century, Funk Island was being mined of its auks, mainly by New Englanders. In 1775, Newfoundlanders petitioned Britain to restrain the slaughter. Some Yankees who were caught skinning birds and plundering eggs on Funk were brought to St. John's and publicly flogged. The controversy finally provoked a legal prohibition of auk-skinning and feather-taking in 1794, but the poaching continued. By the first years of the nineteenth century, Funk Island was barren of auks.

The auk's doom came on June 3, 1844. That morning, Keil Ketilsson, Jon Brandsson, and Sigurdur Islefsson set out from the Icelandic fishing village of Stadur for the rocky, volcanic islet of Eldey. They left in search of auk eggs and feathers, which by then were so rare they commanded fabulous prices from European collectors. Eldey had once been thick with auks in the summer breeding season, but all that was left on that day was a nesting pair. The fishermen killed them. One egg lay on a lava slab. It broke at Keil Ketilsson's feet, and that was the last of the great auk.

The case of the Bermuda petrel is a strange one. In 1616, Bermuda's colonial officials passed a law to protect the island's cahows, as Bermuda petrels were known then. The small seabirds knew no predators in their nesting colonies until the sixteenth century, when mariners—and, more fatally, their egg-eating pigs, cats, and dogs—began winnowing Bermuda's wild bird populations. A stricter law was adopted in 1621, and the cahows were saved, but nobody knew that until the 1950s. That was when Bermuda's petrels, long written off as extinct, were rediscovered.

Even the story of the passenger pigeon, so often told as a simple tale of barbaric human stupidity, is not without its heroes. Though they were extinct by 1912, North America's passenger pigeons were probably the most abundant bird species on earth in the early years of the nineteenth century. The Hurons saw the souls of all their dead in those great flocks, and among North American settlers, there was nothing irrational or bloody-minded in the common belief that any kind of hunting restraint was silly, that nothing could harm their numbers. The birds massed in nesting colonies over hundreds of square kilometres, and their harvest provided a bounty of cheap food both to the working poor of the growing cities along the U.S. eastern seaboard, as well as to foxes, wolves, lynx, hawks, and eagles.

Passenger pigeons were lovely birds, with blue-grey wing feathers, red breasts, and purple iridescent neck feathers. They were much bigger than mourning doves, and every bit as plump and tasty, and so by the late 1870s their numbers were already dwindling rapidly. The initial decline can be confidently laid at the feet of the wild-game industry, which supported thousands of hunters, shippers, middlemen, and retailers. In a largely forgotten chapter of North American history, wild game was a significant part of the urban diet, well into the twentieth century. Venison, wild trout, rabbit, wild goose, prairie chicken, buffalo, wild turkey, snipe, plover, and a wide variety of shorebird and duck species were common North American dinner fare. As the abundance of so many game animals dwindled, market hunters increasingly turned to passenger pigeons. Conservationists responded, fighting for and winning a series of protective laws, aimed mainly at keeping hunters away from key roosting areas during the breeding season. But the laws proved almost impossible to enforce, partly because so many people thought the laws were unnecessary and not worth obeying.

One of the last great battles for the passenger pigeon occurred in 1878, in and around a massive roosting of the birds in the woods near Petoskey, Michigan. The limbs of the trees lay thick and heavy with passenger pigeons over a vast area, roughly 16 kilometres wide and 60 kilometres long. Dozens of volunteers from the Saginaw and Bay City game protection clubs faced heavily armed hunters. The conservationists were led by Saginaw college professor H.B. Roney, whose volunteers harried the hunters and whose lawyers skirmished with game-industry officials. When it looked like the conservationists were losing, Roney spread rumours that the Petoskey pigeons had been eating poisoned berries. It was a mischievous tactic, echoed more than a century later in the spiking of old-growth cedars with nails big enough to break chainsaws and scare away loggers. But it worked, at least for a time.

The Petoskey battles were repeated elsewhere, on a less dramatic scale, but by the late 1880s the passenger pigeons were doomed. As Christopher Cokinos shows in his heart-rending *Hope Is the Thing with Feathers,* neither side of the debate fully understood just how vulnerable the massive flocks had become. With breeding pairs that cared for only a single egg every year, the productivity of the huge flocks depended upon the safety of their sheer numbers. As the flocks declined in size, the numbers of hatchlings fell exponentially, like compound interest, only in reverse. In the pigeons' final years, the sound of a few shotgun blasts was enough to disrupt an entire breeding season. And as the flocks grew ever smaller, the legions of game hunters that had grown dependent upon them became ever more desperate. All the best intentions, and all the conservation measures aimed to protect the pigeons, proved too little, too late.

The point is, though, that the intention was there. In 1708, several New York counties were already enforcing annual closed seasons to protect turkeys, grouse, heath hen, and quail, and by the end of the eighteenth century, game-protection laws were common in the United States. Even the buffalo had its champions. The story of the U.S. government's deliberate extirpation of the buffalo from the prairies—a strategy aimed partly at subduing the Plains tribes—is well known. What is not so well known is that in 1874, the U.S. Congress adopted a law to prevent the "useless slaughter" of the buffalo, but the law was stymied by U.S. president Ulysses S. Grant. In Canada, the buffalo's demise was widely understood for the tragedy it was, and valiant efforts were made to domesticate the animal as a way of maintaining its numbers. Across the Canadian west in the late nineteenth century, game-protection societies were established to conserve a wide variety of wild species. Even in the American west, Idaho was passing laws to protect buffalo, deer, elk, antelope, and other animals as early as 1864. Wyoming prohibited any killing of buffalo in 1890, and seven years later Montana made the killing of buffalo a felony offence, punishable by a two-year prison term.

By the late nineteenth century, inspired partly by early leaders such as H.B. Roney, the bird conservation movement in the United States was going strong. Led mainly by the Audubon Society, it drew many of its most energetic recruits from the ranks of America's children, who made up the majority of the 50,000 people who signed a pledge in the late 1880s to refrain from killing wild birds. The children appear to have been particularly distressed by the fashion in women's hats, some of which displayed not just feathers but entire stuffed birds.

More than a century after women's taste in hats was threatening the extinction of North America's egrets, terns, spoonbills, and ibises, many creatures are still being pushed to the edge of the abyss by the fashion industry. One such animal is the Tibetan antelope, hunted so vigorously for its fine wool that it is teetering on the brink of extinction, a process exacerbated by the rage for shahtoosh shawls unwittingly set off during the late 1990s by Queen Elizabeth II and fashion model Christie Brinkley.

Just as the trade in rare parrots has led to their rapid disappearance in recent years, the trade in bird feathers was one of the main reasons the skies were growing so silent during the late nineteenth century. Yet it's too often forgotten that this tragedy provoked a powerful response among ordinary people. Early conservationists built a movement that touched off the same kinds of social division, high drama, and political intrigue that marked the great labour struggles of the 1930s and the civil rights movement of the early 1960s. After Audubon Society warden Guy Bradley was murdered in 1905 while protecting birds in Florida, public sentiment began to harden against market hunters.

Americans were losing their patience, and it was the fate of a parrot that touched their hearts.

The last Carolina parakeet died in a Cincinnati zoo in 1918. It was the only parrot species that made its home solely in North America. Much prized for its feathers, the bird was formerly abundant in the forests of the southern United States and came to be regarded by farmers as a pest. Carolina parakeets displayed the endearing but fatal habit of flocking around their fallen comrades, a trait that made them particularly vulnerable to slaughter. By the time of their extinction, the U.S. courts were being harnessed in defence of wild birds.

A landmark case began with the 1919 arrest of Missouri attorney general Frank McAllister on charges of killing 66 ducks out of season. By upholding his conviction, the courts confirmed the federal power to protect migratory birds and resolved legal uncertainties about the force and effect of the 1918 Migratory Birds Convention between Canada and the United States. Justice Oliver Wendell Holmes, the famous American jurist perhaps best remembered for upholding the rights of workers to organize labour unions, wrote the key decision in the case. As a young man, Holmes added his own signature to the children's pledge to refrain from bird-killing.

While its depredations should not be ignored, focusing too closely on hunting can allow more subtle and lethal causes of extinction to go unnoticed, such as in the case of the passenger pigeon. Throughout the late nineteenth century, the great oak, beech, and chestnut forests of the U.S. northeast and midwest were shrinking. The situation caused by the felling of beech trees was especially dire. It takes a beech tree 40 years to produce seed, which was a staple of the pigeons' diet, and the big old beech trees were disappearing the fastest. As a line on a graph, the decline of the passenger pigeon is the same as the line marking the rate at which America's great eastern forests were being felled. The pigeons disappeared so quickly that for several years after the American Ornithologists' Union declared the birds extinct in 1912, many Americans still refused to believe it was possible. Even if they were gone, the thinking went, it must have been because the birds had migrated to South America or had succumbed to some catastrophic viral disease.

That's one of the more peculiar things about extinctions: when an animal disappears—especially a bird—nobody wants to believe it. It's as though the animal simply refuses to leave the human imagination. It becomes a ghost that haunts the sleep of its former tormentors and would-be saviours alike.

The eskimo curlew vanished in this way, not so much with a bang from a hunter's rifle as with a chorus of whimpers. Although the last bird was killed in the 1960s, people were still hearing that chorus in the distant sky as late as the 1980s. Like the passenger pigeon, flocks of eskimo curlews formed like clouds, and its disappearance was also the result of a series of unintended consequences. Hunters were certainly among the culprits—as passenger pigeons became more scarce, bird hunters turned increasingly to shorebirds like eskimo curlews. But habitat loss inflicted deeper wounds.

As the great prairies were emptied of buffalo and converted to wheatfields, cornfields, and cow pastures, an ecological domino effect that ecologists call a *trophic cascade* was set in motion. An especially important domino in the line behind the eskimo curlew was yet another sky-darkening creature, the Rocky Mountain locust. In their final, superabundant years, locusts left everything they touched black and scorched and dead in massive tracts of landscape between the Gulf of Mexico and Saskatchewan. In one especially gruesome swarm, in 1875, locusts devoured every green and living thing over an area almost three thousand kilometres long and 200 kilometres wide. Then they turned the centre of the very continent into a rank and fetid mass of their own dead bodies. A quarter of a century after that event, the locusts mysteriously disappeared from the face of the earth. The last time anyone saw a living Rocky Mountain locust was in 1902.

While one might be forgiven for caring little that the locust is gone, its departure was probably the final nail in the eskimo curlew's coffin, the event that condemned the bird to a state of latent extinction that ended with its oblivion. The locusts had been a primary food source for the eskimo curlews during their long northward migrations from Patagonia to their breeding grounds in Canada's Northwest Territories.

By the early 1900s, an eskimo curlew sighting was already a rare event. The last one seen in the wild was shot in Barbados in 1963. Just as the IUCN was reluctant to concede the extinction of the glaucous macaw even though one had not been seen since the 1960s, so the U.S. Endangered Species Act still chooses to list the eskimo curlew as endangered. Even the venerable Audubon Society hasn't given up hope that there may still be a small flock out there somewhere, unnoticed because of the birds' resemblance to other curlew species. In the 1990s, sightings were still occasionally reported, as though the birds' ghosts were appearing to people from time to time.

In this same way the Lord God Bird, as the ivory-billed woodpecker was called, slowly retreated from the known world into the deep recesses of the human subconscious.

Once abundant in the old-growth forests of the southern United States, the ivory-billed woodpecker was a gorgeous creature, the "king of the woodpeckers," much loved by bird fanciers. For decades, the official verdict had been that the last reputable sighting was of a lone female in Louisiana, in 1944, in a forest on the Tensas River, one of the last forests of the kind the birds require to survive. Wildlife artist Don Eckelberry spent two weeks catching occasional glimpses of the bird for his sketches, while German prisoners of war felled trees all around him and the Chicago Mill and Lumber Company only barely tolerated his presence in the woods. Despite appeals from the U.S. federal government and entreaties from the governors of Louisiana, Mississippi, Tennessee, and Arkansas, the logging company kept right on cutting trees.

More than a half-century after Eckelberry's visit to the Tensas, ivory-billed woodpecker sightings were still being reported to wildlife authorities, universities, and the Audubon Society. They came from forests throughout the southern United States, and from as far away as Cuba. Well into the 1990s, government wildlife officials were still getting calls from people claiming to have seen a Lord God Bird, or to have heard one. The call of the ivory-bill was haunting. Its cry, when injured, was like that of a human child.

The macaws had come from the west, and they'd flown low, down at Curú's beautiful curving beach, right past the wildlife refuge headquarters, which is what they call the cookhouse, a gift shop, and a collection of shacks where research volunteers and visiting biologists stay. The birds had flown all along the beach at about tree height, and they were beautiful and bold and graceful. Then they'd quickly disappeared into the palms just below the *mirador,* a lookout point that commands a grand view of Curú's beach, the Tortugas Islands just offshore, and the blue Pacific beyond.

So I was told, anyway. It was a fleeting event and I'd missed it. At the time, I was just a few paces from the beach. An extended family of howler monkeys had been engaged in difficult and comical negotiations with a group of capuchin monkeys about how to fairly divide a clutch of plantains. I'd found it all fascinating, and while I was taking it in I'd missed the macaws.

It was also because of howler monkeys that I was finding it so hard to stay awake at the observation post, waiting for the macaws to come for their sunflower seeds. The night before, a troop of howler monkeys had roosted in the trees directly above the little roadside inn where I was staying near Cabo Blanco. They'd kept me awake most of the night. There is no charitable way to describe the sound a howler monkey makes. They are small monkeys—an adult male is barely the size of an 18-month-old child—but they have hollow hyoid bones in their throats that act like echo chambers, and when they're giving out of themselves the unearthly bellowing can be heard several kilometres away. They carry on this way to mark their territory, for hours on end. The racket provides much amusement for the surfers who flock down the bone-jarring roads to the remote beaches along Nicoya's southern shore every winter. But when the howlers are at it in a tree right above you on a moonless night, you'd think you were surrounded by giant gorillas, and they were all shouting at you.

So I sat in my camp chair with my borrowed binoculars and a clear view of the two seed boxes through the trees, and waited. I made a great effort to stay awake so I could know for myself the experience of seeing a scarlet macaw where it belongs, in its own forests, with its unmistakable feathered costume of bright red and brilliant yellow flashing and its long, blue-green tail feathers. I watched and waited, and I listened, too, but in the dull, distant chatter of bird calls I couldn't make out the macaw's distinctively resonant squawking, no matter how hard I tried. There were instead the timorous cries of what I took to be parakeets, maybe red-lored parrots, or perhaps yellow-naped parrots or orange-fronted parakeets. Every so often, I would see something flicker from the corner of my eye, only to turn out to be a grackle swooping down to land on one of the seed boxes to peck a few seeds before flying off.

An iguana climbed one of the poles in the clearing where the seed boxes were hanging, and it sat motionless and barely noticeable, sunning itself. The crickets picked up again. After a while, nothing moved. Another coconut fell. I waited, and the lazy sun glinted through the green palm canopy, and the iguanas went quietly to and from their appointments. The strange blue butterflies fluttered through the perfumed air.

While I was waiting for the macaws, a team of 50 ornithologists and field biologists was reviewing the results of a top-secret survey of the Cache River and White River wildlife refuges in Arkansas's Big Woods area. Led by the Nature Conservancy and the venerable Cornell University Laboratory of Ornithology, the scientists were gathering evidence to confirm one of the most credible reports of a Lord God Bird sighting in the 60 years that had passed since Eckelberry's days on the Tensas River.

Gene Sparling, of Hot Springs, Arkansas, had convinced Cornell that a bird he spotted on the Cache River on February 11, 2004, just might have been an ivory-billed woodpecker. Two weeks later, Sparling brought Tim Gallagher, editor of Cornell's *Living Bird* magazine, and Bobby Harrison, associate professor at Alabama's Huntsville College, to the bayou where he saw the bird. A huge black and white bird answering every description of an ivory-billed woodpecker flew across the bayou not more than 25 metres from them. The trio agreed that there was no way they could be mistaken. Harrison sat down on a log and wept.

The Cornell ornithologists and the Nature Conservancy followed up with an intensive search. The result was several fleeting sightings, audiotape of some very woodpecker-like sounds, and, most dramatically, a few seconds of videotape that shows a woodpecker that looks very much like a Lord God Bird. The discovery was announced in April 2005. It was one of the most widely and enthusiastically reported news stories of the year—the ornithological equivalent of discovering a sasquatch, or finding Elvis alive after all.

By July the findings were mired in controversy. Scientists with Yale University and the University of Kansas were the main skeptics, and their dissent was reported almost as widely as the initial story. But the fuss was short-lived. After listening to recordings unavailable during the initial flurry of announcements, the skeptics came on side. Yale's Richard O. Prum said the tapes left him "strongly convinced" of at least a pair of Lord God Birds in the Arkansas swamps. After hearing the audiotapes, Mark B. Robbins of the University of Kansas agreed. He said he was "absolutely stunned."

But unlike those Cornell ornithologists, or those stubborn birders who bushwack across the Volga delta looking for curly pelicans, I spent my last day at Curú resigned to the idea that I would not see the birds I'd come for. I spent the time down at the beach with my wife, Yvette, who was itching to get to the surfing beaches on Nicoya's west coast before we headed home. I'd been told there was a slim chance that I'd see a macaw along the banks of a lazy little river that winds through a palm forest to the beach at Curú, so on our way out we waded up the river, hip deep, as far as a rustic-looking rope footbridge. When we scrambled up the bank, we noticed a sign: Beware Crocodiles. It had not been my best day.

Walking the trail back to the cookhouse, I took comfort in the knowledge that Costa Rica was evidence enough that all roads do not lead into the Valley of the Black Pig, and that the fate of the rare things of the world is not necessarily extinction or the living death of zoos and aviaries. Life's greatest enchantments are those things you have to wait for. You have to persevere, and sometimes you just get lucky. I was telling myself these things on the trail when something caught my eye—a flash of red, high in the forest canopy.

There, in the fronds of a palm tree, were two impossibly beautiful scarlet macaws.

A fish

The Last Giants in the River of the Black Dragon

We've been bewitched by countless lies,
by azure images of ice,
by false promises of open sky and sea,
and rescued by a God we don't believe.
—Yevgeny Yevtushenko, "Ballad about False Beacons"

It had been a cramped and creaky two-hour flight on a 174-passenger Ilyushin 62-M airliner from Seoul, South Korea, and even before the plane landed on the cracked and buckling runway at Khabarovsk Airport, the great unravelling of the Russian Far East was making itself obvious. As the Ilyushin banked to port on its descent, the sunset streamed through the windows and turned the passenger compartment an exceptionally brilliant blood red. It was because of a vast prism of smoke that rose on the horizon, caused by runaway forest fires from all the illegal logging operations hacking their way across the taiga. During the 12 months prior to the lovely sunset that greeted Dalavia Airways Flight 301 on its arrival in Russia, an amount of timber worth an estimated $120 million U.S. had been stolen from the forests around Khabarovsk.

As the plane came to a stop on the tarmac in front of the crumbling terminal, it was surrounded by soldiers in tattered and ill-fitting Soviet-era uniforms. They herded us into long lines on the runway while platoons of nurses ran toward us with armloads of gigantic thermometers, one for each passenger. The nurses gestured to indicate that we were to stick the thermometers' huge, cold metal tips into our armpits, and so we all stood in this absurd way, in queues that snaked into the airport's arrivals section. It was all in aid of showing some sign of vigilance following a SARS outbreak in China's Heilongjiang province, just a few kilometres away. After some chaos at a series of improvised medical stations, the passengers were hurried in the direction of the Russian customs desks.

While I was watching the ground crews out on the tarmac tossing our luggage from the cargo hold into the back of a rusted army truck with a big red star on the side of it, a wiry and frantic-looking man walked by. He was wearing a baseball cap from the St. Peter's Fly Shop in Fort Collins, Colorado, so he caught my eye. It turned out to be Misha Skopets, one of only a handful of people I'd ever met who lived anywhere near this part of the world. Misha lived 1500 kilometres northeast, in Magadan, on the Sea of Okhotsk. It turned out that he'd just happened to be in Khabarovsk, but he intended to stick around for the symposium on fisheries conservation that had brought me there. This was great luck.

I'd met Misha only once, a year before, at an international conference on salmon in Portland, Oregon, but he'd left an impression. Misha's like that. The 47-year-old fisheries biologist has been called the Indiana Jones of Russian fisheries conservation. Few rivers in the

Russian Far East are unknown to him. He has discovered several fish species, one of which—a salmonid—was so unusual that it warranted a completely new genus. He'd found it at the bottom of a subarctic lake inside the remote El'gygytgyn crater, in Chukotka. Misha is good company, and I was pleased to see him, and so were several of my fellow passengers from Flight 301. About a half-dozen other North Americans had come for the symposium, and most of them had at least heard about Misha.

The year before, in Portland, Misha had delivered a sad but blistering account of what Russia's "shock therapy" transition from communism had meant for the country's Far Eastern rivers and for the marine ecology of its Pacific waters. I'd come to the Khabarovsk symposium mainly because the attendance list included just about everyone who would know about the great unravelling Misha had described. This isn't to say the tragedy is known only among a coterie of experts. It's just that the magnitude of what has been happening is largely unknown outside of Russia.

The day I arrived in Khabarovsk, there was an article on the front page of the *Pacific Ocean Star,* Khabarovsk's daily newspaper, about a police investigation into the assassination of Valentyn Tsvetkov, Magidan's state governor. The investigators were saying that Tsvetkov had been killed by the organized-crime bosses who had gained control of whole sections of the coastal fishing industry. But no one was certain, and nor could anyone say with any certainty who had murdered Major General Vitaly Gamov, the senior fisheries enforcement officer on Sakhalin Island. The year before, Gamov had begun a crackdown on high-seas poaching. Some said his assassins were from a Japanese organized-crime syndicate. Others pointed to the Japanese gangs' Russian-mafia clients. It was a mystery.

After we'd collected our luggage and made it through customs, we stood around waiting for our buses into town and the talk turned to a scandal within Khabarovsk's community of conservationists, fisheries bureaucrats, biologists, and fishing guides. A series of photographs had been making their way around town. In the pictures, a well-known "new capitalist" from Khabarovsk, an avid sportsfisherman, was shown standing on a snow-covered riverbank, posing for the camera, holding up a giant, dead Amur River taimen.

Taimen are a kind of salmon, from the ancient genus *Hucho,* a race of giants that arose about 40 million years ago. They are known as the "tiger of ichthyofauna" because of their size and ferocity, and also because they were becoming as rare as tigers. The Amur taimen are the world's largest salmonids: an adult specimen looks like an Atlantic salmon but is the size of a full-grown man. Taimen have been known to weigh 100 kilograms, and to reach lengths of two metres or more. Tribal fishermen along the Amur were known to bait their hooks with dead dogs to catch them, but nobody was catching taimen very often anymore, at least not the big ones. The Amur's legendary taimen are rapidly disappearing, becoming as scarce as the Amur tiger, the largest cat species on earth, whose numbers have fallen below 500. It is mainly because of poaching.

What had got people talking about the photographs was that the fish was especially huge and it was dead. A broadly supported campaign had sprung up along the Amur to stop the killing of taimen, to allow only careful catch-and-release sportsfishing. The new capitalist in the photographs was popular in fishing-guide circles in Khabarovsk. He should have known better, everybody said, and showing off like that, with photographs, wasn't helping.

The Amur River is one of the ten great rivers of the world. It is also known by its more ancient name, the Heilongjiang, which means River of the Black Dragon. It has always been famous for its mysterious giant fish. Within its broad main stem and its tributaries, along with the taimen, is the Amur kaluga, the largest freshwater fish in the world. Found nowhere else on earth, the kaluga is a kind of sturgeon that can reach the weight of a dozen men, 1000 kilograms, and grow to six metres. Both the taimen and the kaluga are revered characters in the mythologies of the Amur's aboriginal peoples: the Nanai, the Nivkh, the Udegei, and the Ul'chi. Taimen sometimes showed up in those stories as werewomen whom young men were prone to fall in love with. Kalugas showed up in stories in different ways. In one famous tale, a boy child, believed to have drowned in the Amur, was later discovered alive, having been raised by a kaluga who served as his stepfather.

Like the taimen, the kaluga was rapidly disappearing into the boiling cauldron of gangsterism and poverty that had made fish poaching, illegal hunting, and illegal logging the dominant industries in much of the hinterland of the Khabarovsk territory. It was a state of affairs that was making a lot of people very angry and positively nostalgic for the days before perestroika. It was also a central part of an important story rarely told in the triumphalist West, where the received wisdom was that the Russians were perfectly happy to have thrown off their totalitarian yoke. You didn't have to spend much time in Khabarovsk to figure out that the situation was a lot more complicated than that.

More than a decade after the close of the Soviet era, the City of Khabarovsk was a decaying but strangely genteel place. Along its tree-lined boulevards were huge tracts of ramshackle apartment blocks and loghouse-style frontier buildings with onion-domed roofs, all jumbled together on a series of hills overlooking a broad reach of the Amur River. The city is the capital of the sprawling territory of Khabarovsk Krai and home to about 750,000 people. It's closer to Vancouver than to Moscow and lies on almost exactly the same latitude, so its climate is more forgiving than that of most Russian cities. They say Khabarovsk is at its best in the early spring, during the week that begins with the old Soviet high holy day of May Day and ends with Victory Day, which commemorates the Great Patriotic War, as Russians call the Second World War. Most people try to take the week off; they trundle to and from their dachas and vegetable plots out in the countryside, and most schools and universities are closed. There are concerts and marches and parades, and there's a good deal of vodka-drinking involved. As it happened, we'd arrived in Khabarovsk that very week.

Although Khabarovsk is a city of dreary Stalin-era office complexes, monuments to Bolshevik martyrs, and dirt-road ghettos where the Gypsies live, it is also a city of grand old Orthodox cathedrals, sprawling parks, museums, and galleries. A statue of V.I. Lenin still gazes out over Lenin Square, and city hall is still on Karl Marx Street, but there is a liveliness about the place. This is partly because of the city's multiethnic hodgepodge, and partly because 19 colleges, institutes, and universities are only a short, rickety tram ride from downtown. It's a bit of Prague in the 1960s, and also a bit like Chicago in the 1930s. There are sidewalk cafés everywhere, and college students engaged in animated conversation, but there are also hordes of grim-faced young men in crewcuts and black leather jackets, loitering on street corners or climbing in and out of flashy black limousines with smoked-glass windows.

During May Day week, the downtown streets were alive with buskers playing accordions and violins. Hawkers plied their trade at vegetable stalls and impromptu kiosks, selling milk and soft drinks and chocolate bars, and the magpies were building their nests in the trees. Children played in the parks and lined up in disorderly queues for reindeer rides and pony rides. And one morning in the Central Hotel, a crackerbox palace of crumbling plaster and broken elevators overlooking Lenin Square, I was awakened at dawn by a deafening noise. It turned out to be the sound of dozens of Russian tanks, troop carriers, and mobile missile launchers streaming through the city's streets. Columns of soldiers were marching in their thousands just below my hotel-room window, pouring into Lenin Square. It was all a bit comical, but sad, too. Throughout the day, thousands of people stood in dead silence, watching proudly as battalions of soldiers paraded around the square. When the bands played the old Soviet anthem, legions of teary-eyed old men stood to attention in overcoats decorated with tarnished Communist Party medals. All day, women stood in a silent vigil around the square, holding aloft their old portraits of Khrushchev, Brezhnev, and Stalin.

The nostalgia made some sense. For the vast majority of Russians, the radical disassembly of the Soviet system had been a disaster. In the Russian Far East, the collapse of order had taken on a distinctive, wild-west character. The picture that emerged from the testimony of more than 40 senior government officials, fisheries bureaucrats, enforcement officers, biologists, conservationists, and aboriginal leaders at the symposium I'd come to attend was, if anything, worse than the portrait Misha had painted in Portland the year before. Sergei Zolotukhin, the Russian government's sober and heavy-browed senior salmon biologist for the Khabarovsk territory, offered a grim forecast over lunch one day. Without a radical change in the entire political and economic regime, the situation was pretty well hopeless, he reckoned. "It is always poaching now," he said. "The law has no teeth."

How things had gotten so bad so quickly is a story that begins with the rapid emergence of a criminal oligarchy among the higher ranks of the old Soviet *nomenklatura* after Boris Yeltsin seized state power in 1991. Yeltsin immediately put Russia on a "markets first" road, and within a decade, criminal enterprises dominated the oil industry, the real estate industry, the wholesale food business, the liquor trade, the aluminum industry, the hotels, and the restaurants. By 1997, the CIA quietly reported that about 40 percent of Russia's economy had fallen into the hands of about a half a million of Russia's 145 million people in a network of organized crime syndicates. The gangsters ran about half of Russia's 25 largest banks. The amount of currency that left Russia illegally during the 1990s has been variously estimated at $220 billion to $450 billion U.S. Entire industries were liquidated, the profits were reinvested outside the country, and millions of workers were suddenly jobless. The entire country was looted.

Russia's annual gross domestic product fell by half during Yeltsin's regime. In the first half of the 1990s, hyperinflation wiped out all but about 1 percent of the savings and pensions of ordinary Russians. In their sad attempts to recover their losses, the people succumbed to a mania of investment funds. During the second half of the 1990s, the funds proved no better than elaborate pyramid schemes, and they collapsed upon millions of investors, one by one. Life expectancy among Russian males fell by seven years, to 57, roughly the same as in Sudan. The wave of premature deaths was matched by a rise in infant mortality. A demographic implosion spread throughout the country, and the population actually shrank by several million. The country had seen nothing like it since the Stalinist terror of 1937–38.

The situation in the Russian Far East was as bad as anywhere. Covering an area more than half the size of the European continent, the Russian Far East looks out on the Bering Sea and the Sea of Okhotsk in the north and the Tartary Straits and the Sea of Japan in the south. It includes the huge Kamchatka Peninsula and Sakhalin Island, and from west to east it runs from the Mongolian frontier to a point only a few sea miles from Alaska. The largest of the region's ten territories is Khabarovsk Krai, which wraps around the northeastern tip of Manchuria and sprawls out to the north as far as Yakutia. About half of the Russian Far East's seven million people live in Khabarovsk Krai and in Primorye Krai, Khabarovsk's neighbour on the southeast. Primorye forms a north–south panhandle that takes in the rugged Sikhote Alin mountain range between China and the Sea of Japan. On its southern tip, the grim harbour town of Vladivostok, the region's second-largest city after Khabarovsk, looks out on the ocean just a few sea miles from North Korea.

In the late 1970s, the Russian Far East was a place of great promise. It was precisely the region's back-of-beyond appeal that had drawn

Misha Skopets from what could have been a perfectly boring life back in Yekaterinburg, in the Ural Mountains. When Misha set out for the Russian Pacific, he had high hopes, an enthusiasm he would later attribute to the Jack London novels he'd read when he was younger. Things were looking even brighter in the late 1980s, the first years of perestroika. Misha earned his doctoral degree in biology in Vladivostok, and his career trajectory took him on long journeys into the region's hinterland. Life was good. Then things started falling apart. The largest economic enterprises in the Russian Far East had been Soviet collectives, but during the 1990s most of them became private monopolies controlled by former Communist Party administrators, who had transformed themselves, almost overnight, into "new capitalists." Private businesses ended up with almost all the forest licences and most of the region's vast mineral deposits. Unemployment soared, wages were slashed, and workers routinely went months between paycheques. The forests that remained in public hands were being clear-cut by criminal outfits. Journalists who tried to report on the situation did so at the risk of their lives.

In 1994, armed thugs burst into Vladivostok's PKTV television station, shot and killed a technician, and destroyed the broadcasting equipment. Newspaper and radio reporters were routinely beaten and threatened. When Vladivostok's citizens elected a populist anti-corruption mayor, the state government sent a platoon of 200 police to city hall, where they broke down doors, evicted the mayor, and chased 120 senior civic officials out of the building. In response to the crisis, Russian president Boris Yeltsin appointed a new mayor—one nominated by the local crime bosses. By 1996, gangsters ruled the streets of Vladivostok, and living conditions were worse than they were even in the darkest days of the Great Patriotic War.

Russia's Pacific fish stocks were plundered. Fishing quotas were assigned to new quasi-legal corporations that sublet their shares to foreign fleets, mainly shady Japanese enterprises that counted, as a simple cost of doing business, the lease fees they deposited into gangsters' foreign accounts. Russian naval and coast guard forces were powerless to stop what had degenerated into a massive poaching free-for-all. The naval base at Russky Island, which lies almost within sight of Vladivostok Harbour, was cut off from its basic food supplies; more than a thousand officers and crew had to be evacuated to the mainland to be treated for severe malnutrition. Four sailors starved to death.

Through the 1990s, Japanese and Korean fishing boats in the North Pacific and the Sea of Okhotsk were scooping up all the pink salmon runs bound for Sakhalin Island and Kamchatka. The salmon that made it through the high-seas fisheries gauntlet into the rivers of the region often fell to poaching syndicates, some of which were so well capitalized that they could afford to build roads to the more remote rivers. Roe-stripping—tearing egg sacks out of female salmon before they have the chance to spawn—became the main source of income in vast areas of the Kamchatka, Magadan, and Khabarovsk territories. The illegal sale of chum salmon roe to Japan became a multi-million-dollar business. The poachers' catch in the Russian Far East ended up eclipsing the legal catch by Russian commercial fishermen.

It was anarchy, Misha said. One sunny afternoon, to illustrate his point, he took me to the Khabarovsk central market, a half-hour's walk from the Institute for Water and Ecological Problems, where the symposium was underway. Half the market was a mayhem of open-air stalls, and the other half was only slightly more organized, inside cavernous buildings. The vendors were Russians and Mongolians, Manchurians and Gypsies. Some had sneaked across the border from China, and others were selling merchandise of dubious origin that they'd brought from as far away as the Caucasus Mountains. Everything was on offer, from motorcycle parts to clothes, plumbing supplies to canned vegetables, cars to school supplies. We spent the afternoon wandering among the fish stalls.

There were tonnes of fish for sale, from dozens of species. Most of the fish had been caught in the 4400-kilometre-long Amur. Because the river traverses both the Monsoon region and the Siberian region of the temperate climatic belt, the Amur is richer in its diversity of species than any other river in Russia. In its waters live 108 species from 79 genera, 23 families, and 10 orders from a bizarre blend of Sino-Indian and Holarctic types, including a whole range of species that occur nowhere else on earth and a dazzling array of relic species from the Pleistocene and Pliocene epochs. Laid out in trays in the Khabarovsk market's stalls were skygazers and pikes, rainbow smelts and river-horses, ciscos and carp, and catfish and barbel chubs. Among them, in their orderly little rows, the tragedy unfolded.

"There, a yellowcheek," Misha said. It was a juvenile, only about the size of a small salmon. It's a barracuda-like fish that can grow to roughly 50 kilograms. "This is red-book," Misha said, referring to Russia's "red list" of endangered species. He lifted it out of the pile, to take a closer look. "Definitely red-book. Very rare fish." He put it back in the tray with the others. "It is a great sports fish, too," he said. "It is fighting like crazy." We continued on around the stalls.

"This one is red-book," he said, pointing to a Chinese perch. We passed by pink salmon, chum salmon, and Pacific cod, "just like British Columbia," Misha said, and there were rockfish from Sakhalin Island, and flounder from the Sea of Okhotsk. We saw bigheads, burbots, and lenoks, then a Mongolian redfin, and a huge chunk of flesh that Misha pointed at and said, "kaluga." He shook his head. "It is very sad shape now. It is red-book." And then there was another big piece of flesh, with slightly different Cyrillic script on the piece of paper in front of it. "Amur sturgeon," Misha said, meaning the distant cousin of the kaluga that can reach three metres in length. They're found only in the Amur, nowhere else on earth. "It is also red-book." Then, among some saffron cod and black halibut and Amur pike, there was a black breem. "That is red-book also." We wandered through the crowded fish market a while more. "And that's just what you can see," Misha said. "Ask any of these people, and they will get you any kind of caviar, or taimen, or anything you want."

The grim reality of the Khabarovsk fish market played itself out over the following days in a dimly lit seminar room at the Institute for Water and Ecological Problems, a long brisk walk down Karl Marx Street from the Central Hotel. Ostensibly, the conference delegates had gathered to discuss whether the relatively pristine areas remaining in the Russian Far East should be set aside entirely and permanently, inside heavily patrolled "protected areas," to preserve at least a representative fraction of the region's fish stocks, especially salmon. The idea was being pioneered by the Wild Salmon Center in Portland, Oregon, which had been generously supporting the work of Russian fisheries scientists in the wake of the Soviet Union's collapse.

The protected-area strategy was a bit of a doomsday scenario, based as it was on the assumption that the situation was indeed as hopeless as Sergei Zolotukhin, the big-browed government salmon biologist, made it out to be. But Russia was not Costa Rica, where a stable and relatively comfortable society had invested its hopes for the future in ecological restoration and a network of protected forests. Even in North America, the strategy of setting aside large parks and wilderness areas had not been a raging success, as William Newmark's research showed in the "Night of the Living Dead" chapter. "Faunal collapse" followed the creation of too few parks, too small and too isolated from one another.

Still, for the Russians, the idea had some gravity. For many, the establishment of protected areas, especially for salmon, was the only hope, so long as local people supported the zones. Obviously a lot more money could be made in ecotourism and in catering to well-heeled foreign anglers than in mining the region's salmon and sturgeon and taimen into extinction. But it was a complicated subject. Opinion was divided.

Sergei Makeev, chairman of the Sakhalin Wild Nature Fund and a senior fisheries biologist with the Sakhalin territorial fisheries department, argued there was a danger that salmon-protected areas would end up as special fishing preserves for the "elites." Most people in the region had seen quite enough of that sort of thing, Makeev said, and without strong local support, a protected-area approach would fail.

Zolotukhin, meanwhile, proposed that the more remote salmon rivers be set aside for private tourism companies on long-term leases, in a quasi-privatization arrangement. That way, the companies, flush with hard, foreign currency, would have the motive and the wherewithal to control abuse. They could restrain poaching and support scientific study at the same time, he said.

Misha's view was that there was no point in even trying the approach except in the most remote areas, and only if it provided some tangible benefits to the local people, particularly the "small nations," as Russians call aboriginal peoples. The situation was actually worse than Zolotukhin was letting on, Misha said. Most of Zolotukhin's enforcement staff lived in dire poverty, counting themselves lucky to earn the equivalent of $150 U.S. a month, and there weren't enough of them to control poaching anyway.

Most delegates were careful to point out that they did not mourn the demise of the Soviet system, but speaker after speaker described the desperation of the new epoch. "It has been like letting animals out of a zoo," Misha said. "They have to find food for themselves somehow now. Everybody wants to earn some living. But you can't, unless you break the law."

Few placed much hope in emerging democracy, at least not the kind they were seeing. What was needed was some way to restore order to the region, but Moscow got what Moscow wanted, everybody said. It certainly didn't seem to matter what the people of the Russian Far East wanted. In an in-depth public opinion survey the Khabarovsk Wildlife Foundation conducted in 2002, the people of the territory showed deep concern about the ecological damage in the region. They were overwhelmingly in favour of greater environmental protection, and were particularly concerned about the unique forests of the Sikhote Alin mountains, where "protected areas" were turning out to be protected in name only.

The Sikhote Alin mountain range in Primorye supports possibly the strangest forests on earth, and they are especially at risk. Because of the mountains' unique diversity, the United Nations Educational, Scientific and Cultural Organization (UNESCO) declared the central Sikhote Alin range a World Heritage Site in 2001. The mountains were never glaciated during the Pleistocene ice ages, so the forests are a mix of taiga and subtropical. Amur tigers and Himalayan black bears inhabit the same forests as reindeer, wild boar, and one of the world's rarest mammals, the Far Eastern leopard. But the Sikhote Alin had become one of the most endangered ecosystems on the planet, and the people wanted more and better wilderness reserves. In the Khabarovsk survey they demanded a crackdown on illegal logging and poaching and stricter wildlife-protection laws, as well as more rigorous and effective enforcement of those laws.

The public values evident in the Khabarovsk survey are remarkably similar to those of Canadians, on the opposite side of the Pacific. An April 2000 survey by Canada's Habitat Conservation and Stewardship Program showed that a majority of British Columbians want salmon runs protected and salmon habitat conserved, even if it means higher taxes or a slowdown in economic development. Rural and urban British Columbians alike ranked the commercial value of salmon well below the contributions salmon make to the ecological health and "beauty of the region" and to the enhancement of "community involvement."

The Khabarovsk questionnaire respondents were also given an opportunity to record opinions that they thought the survey's questions didn't adequately account for. The cumulative written responses read like a manifesto. They called for a swift crackdown on illegal logging and on the corrupt police that provided "cover" for the operations. They wanted an end to the export of raw logs to Japan and China. They wanted a special reserve zone along the Kopi River, a prohibition of all high-impact activities in specially protected areas, fair wages for conservation staff, a public-awareness campaign for children about the value of natural resources, a revival of the Soviet-era Young Naturalists organization, and an assurance that protected areas would be enjoyed by all people, not just "senior administrators." These opinions came from people of all ages, throughout the territory, and there was no noticeable difference of opinion between urban and rural residents.

But the people weren't getting what they wanted. A new criminal class was in control. Politics had become the hobby of choice for crime lords because elected deputies in the new Russian parliament were immune from prosecution. Votes were routinely put up for public bid. The going price for important votes could run as high as $35,000 U.S., and the entrenched kleptocracy had become ever more immune from popular opinion as the years passed. The oligarchy held power so tightly that no scandal, no matter how outrageous, made any difference.

When Yeltsin handed the reins to Vladimir Putin, the spy-agency strongman who talked a tough line on corruption and organized crime, there was hope in Khabarovsk, Vladivostok, and Petropavlovsk. But it soon dimmed. In May 2000, Putin abolished the 200-year-old Forest Service. Then he abolished the State Committee on Ecology, the equivalent of the U.S. Environmental Protection Agency. He then handed the committee's functions to the Ministry of Natural Resources, which was widely regarded as the main government conduit for the plundering of the Far East's natural wealth. That same year, Putin visited Sakhalin Island to encourage oil companies that wanted to build offshore drilling platforms around the island. He dismissed the concerns of the area's conservation groups, accusing them of being nests of foreign spies. Former Russian prime minister Viktor Chernomyrdin accompanied Putin on the trip to Sakhalin and busied himself shooting bears in a "nature reserve" on the Krilyon Peninsula. Chernomyrdin's conduct was an obscenely graphic illustration of the problem: when protected areas were established at all, they weren't easy to protect, or even to keep.

In the Khabarovsk territory, there were 60 nature reserves. All but 10 existed only on paper. In Primorye, the Golubichnaya River watershed had been legally protected and was fully enclosed by a nature reserve in the Sikhote Alin mountains. In the 1970s, the river was teeming with char, taimen, and a variety of salmonids; by the late 1990s, the poachers had finished it off—the fish were gone. It was the same on the Bolchi River. Shortly after the Russian Far East was opened up to private logging corporations, the Bolchi, on the Primorye coast, was set aside to keep the logging industry out. The point was to save the fragile salmon habitat of the valley. Ten years later, the Bolchi was still pristine, but its spawning beds had been rendered almost barren of salmon, mainly by poachers. In 2001, the Sakhalin government abolished wildlife refuges in the Noglik and Smirnykhovskii districts. Then, in 2002, the reserve on the Krilyon Peninsula was turned over to a private hunting venture, the Sakhalin Krilyon Company. Those three acts of the Sakhalin government alone eliminated 182,000 hectares of protected area.

The same thing happened with fish species protected by Russia's weakened endangered-species law. It didn't even matter if they were protected by international law, through the Convention on International Trade in Endangered Species (CITES). That was a point Zolotukhin himself was prepared to make.

Take the Sakhalin taimen, in the rivers of the Khabarovsk territory, he said. Sakhalin taimen are an extremely rare type of Amur taimen that don't spend their entire lives in rivers but also go to sea. They are giants, too, and so rare and mysterious that the Japanese call them "ghost fish." The numbers of Sakhalin taimen returning to spawn in Khabarovsk's rivers had dwindled so sharply that Zolotukhin's staff reckoned perhaps only 4000 remained. Of course they should be strictly protected by law, Zolotukhin said. But strict protection had been extended to Sakhalin sturgeon, another rare sea-going giant, which was also protected by international law. "What actual protection did such listing provide?" Zolotukhin asked. "None. It is almost disappeared in the coastal rivers."

Zolotukhin certainly wasn't being naive about how successful a network of small protected areas might be. In one of his submissions, he made the point clearly: "In conditions of unstable legal situation ... it is very difficult to expect that establishing a special regime of nature use would limit further robbing of natural resources." It was just that he saw no other way. Save the last of the best, the reasoning went, and leave everything else.

This had a certain grim logic, because even when fisheries enforcement officers managed to do their work, their efforts often came to naught. In 2003, the Amur Fish Authority laid more than a thousand criminal charges against poachers. The overworked prosecutor's office followed up on only 26 cases. Only a handful of those actually ended up in court. Fish inspectors on the Amur earned monthly salaries that amounted to less than $100 U.S. Every night, from May to September, poachers in fleets of rowboats set hooks in the Amur River for kaluga. The caviar from a single fish caught in one night could easily yield a poacher the equivalent of a fish inspector's annual salary. Bribery was commonplace.

Whatever the merits of the idea of protected zones around remote rivers, the approach was based on the grim and largely unspoken assumption that big rivers flowing through heavily populated areas were a lost cause. It was a hard thing for the Russians to accept, but the proposition also wasn't easy to challenge. The Amur River, for instance, had lost about 90 percent of its salmon, mainly during the 1990s. The salmon fed not only people, but bears, hawks, and giant fish like the taimen. More might have been preserved, but for about a thousand kilometres of its length, the Amur forms the border between Russia and China, and when it came to conservation matters, the two countries were barely on speaking terms.

Russian authorities fought hard to convince China to scale back its pollution of the Amur's many Chinese tributaries. Chinese authorities responded by demanding that Russia curtail its illegal fishing. The two countries did form a special Amur fisheries commission in 1994 but then they spent the next decade arguing about quotas and net-mesh sizes. In the lower section of the Amur, which falls entirely within Russia, fishermen caught as many fish from migratory stocks as they could before the fish reached Chinese waters. At times the only thing stopping the Russian fishermen from catching everything in the Amur was the pollution from the river's Chinese tributaries, which made the fish downstream inedible for months at a time.

But poaching was the main reason the great River of the Black Dragon was losing its giants, and all its smaller species, too. That was one thing everybody seemed to agree upon. One day when Misha stood up to make a point about poaching, he put his notes down, set aside the language of environmentalism, and spoke from his heart.

"I think these poachers should be bombed," he said. The hall fell silent.

"These protected areas are fine, but we need to send helicopters, military helicopters. But this does not happen. The Russian Far East is colony of Moscow. In Moscow, they take money from the oil industry, and from these fishing companies. We are an occupied territory."

There was a hush in the hall.

"It is still an empire," he said. "Maybe one day, they will not be able to rule us, these tsars from Moscow. Things are not getting better. They are getting worse."

Later I asked Misha to elaborate. He said he was quite serious. "This is why some people are talking about having true independence in Khabarovsk."

But then he wanted to make a point about what he meant by "poaching," and he reiterated what I'd heard from just about everyone at the symposium. There were poachers, Misha said, and then there were poachers. There were the well-organized criminal gangs in Kamchatka that employed hundreds of people in clandestine roe-stripping expeditions. There were also the quasi-legal coastal fishery enterprises that had won concessions and quotas by bribery and intimidation. But then there were the tens of thousands of poor people in the hinterlands who needed to fish just to survive. Those people were mainly from the small nations, peoples like the Nanai and the Nivkh on the Lower Amur. They were entitled to fish as their forefathers had, Misha said. It's what everyone said.

"It is not so simple," he remarked, and pointed to a tall man in a suit standing in the hallway outside in seminar room. "Him, he is not really a bad man. He was being ignorant. But things do not change overnight."

I didn't know what Misha meant. "It is the man in the photographs," he said, "with the big taimen."

Misha introduced us. The man handed me his business card, an outlandish-looking thing with a picture of an Amur tiger on it. Underneath his name, it said, "Extreme tourism, fishing, ecological tours, photo, video." This was the guy everybody had been talking about at the airport.

Over coffee he pulled two 8-by-10 glossy colour photographs from a little briefcase. "You can have," he said. In one photograph, a man in a red ski suit lies in the snow beside a giant fish that looks just like an Atlantic salmon, except it's as big as the man. In the background are a tent and a pile of camping gear on a gravel bar and an inflatable boat with two fishing rods leaning up against it. In the other photograph, my coffee companion himself is standing in knee-high wading boots, in the same spot, in front of the inflatable boat. He's holding the taimen upright, by its gills, and he's straining under the weight of it. The taimen's tail is bloody, trailing in the snow. He told me he'd caught it at the mouth of the Anui River, at its confluence with the Amur, about 250 kilometres downstream from Khabarovsk. "It is my hobby," he said, smiling. He said he thought the photographs would be good publicity for his guiding business, a sideline to his real estate business. "I can't guarantee such big fish," he said, "but smaller fish, I can guarantee."

It was a difficult conversation. I tried to ask him what it was like, catching the fish, but my questions just confused him. I don't speak Russian, and his English was rudimentary. It was just as well, given what I learned later when I showed the photographs to two Russian taimen experts, Anatoly Semenchenko, co-author of *Taimens and Lenoks of the Russian Far East,* and Igor Parpura, director of the federal nearshore fisheries laboratory in Primorye. They were both mightily impressed with the size of the fish and similarly impressed with the quality of the photographs. But it didn't take Semenchenko and Parpura much close inspection to discern that the fishing rods in the picture were just for show. The prints were clear enough to reveal net marks on the taimen's head, meaning it had been caught illegally. I told Semenchenko and Parpura what he had told me—that he'd caught the fish during the first week of April, a time of year when taimen are still holding in deep cold pools below the ice, waiting for the spring. "That is the other thing that bothers me," Parpura said. "It is obvious they knew this was a wintering pool, and they just went out to get it."

But as Misha had said, there were poachers, and then there were poachers.

Later that week, Tatiana Khetani, chair of the Indigenous Small Nations of the North of Magadan Oblast, described to the symposium how poaching had become a necessary way of life for so many of the small nations of the region. A small, dark woman who could have been Ojibwa or Cree, she took her place at the dais at the front of the room and spoke nervously, quickly, and quietly.

When the aboriginal collective-farm system in Magadan was dismantled in 1994, everyone was thrown out of work, Khetani explained. The amount of fish the aboriginal communities were entitled to catch had been 9500 tonnes a year, but the allocation was cut back, bit by bit, and gradually reallocated to new private businesses. By 2001, the aboriginal allocation had fallen to 486 tonnes, one-twentieth the amount it had been. The Even people, one of Magadan's small nations, started eating the reindeer they had formerly raised for cash, because there was no way to get their meat to market anymore. But they were eating too many reindeer to allow the population to maintain itself: the people started the post-Soviet era with about 100,000 reindeer; 10 years later, the herds had fallen to 17,000. For hundreds of aboriginal families, poaching was the only thing keeping them alive. The unemployment benefit for an out-of-work reindeer herder in Magadan was 120 rubles a month. For a single kilogram of salmon roe, he could earn three times that amount. "Of course, when merchants come to villages with food and suggest trade for caviar, people cannot resist," Khetani said. "When you have nothing to feed your children, you have little choice."

Then there was poaching of another sort altogether. It was the industrial-scale kind.

In an unusually frank assessment prepared for the symposium, Vladimir Belyaev, of Russia's federal fisheries institute in Khabarovsk, and G.I. Sykhomirov, of the regional Institute of Economic Research, explained that though the giants of the Amur River and its tributaries had been vanishing for generations, the really big losses came after Yeltsin seized power, when "the poachers became organized."

The giants' demise started soon after Russian, Manchurian, and Chinese settlers began to flood into the region during the eighteenth and nineteenth centuries. In the dying days of the Romanovs, before the Great October Revolution of 1917, the all-species fisheries harvest from the Amur River sometimes reached a staggering 100,000 tonnes per year. That's the equivalent of the weight of nearly two million people.

Although the Soviet era brought with it many progressive reforms, as often as not central-state planning botched the job. But the Amur's great giants, such as the yellowcheek, the taimen, the sturgeon, and the kaluga, were at least nominally protected. Sturgeon farming, for the caviar industry, lifted the pressure from wild stocks. The Soviets banned kaluga fishing in the Amur in the 1930s.

Even during the darkest police-state tyranny of the Stalin era, there was poaching on the Amur, but the difference between the poachers of that time and those after the Soviet collapse was all about scale, power, and impact, Belyaev and Sykhomirov explained. The new poachers, the industrial-scale poachers, were of an order far more malevolent than anything that had gone before. "They have good transportation and communication equipment, weapons, protection, support from militia, fishing inspection, and other nature protection organizations," their assessment noted. "They have a good network for selling fish all over Russia, and in the countries of Southeastern Asia."

Ten years after communism's collapse, the Amur's all-species biomass had declined to perhaps 1 percent of its former abundance, Belyaev and Sykhomirov concluded. It was an ecological disaster comparable to the Soviets' wilful destruction of the Aral Sea, a horror that has been widely reported outside Russia and often described as the worst environmental disaster in human history. And like the draining of the Aral Sea, the destruction of the Amur took an enormous human toll. On the Amur, the human cost was borne mainly by aboriginal communities.

Like their counterparts on North America's west coast, the aboriginal peoples of the Russian Far East have always been primarily fishing cultures. While the Communist Party initially adopted an enlightened policy toward aboriginal peoples, favouring the persistence of indigenous culture, the later Soviets herded them into collective farms that were, at least at first, every bit as soul-destroying as the worst of North America's Indian reservations. Still, by the late 1960s 168 Nanai, Ul'chi, and Nivkh communities persisted along the Amur. Their collective-farm economies were based on fishing, and an annual allocation of 30,000 tonnes of various kinds of fish provided them a reasonable livelihood. But within ten years of the Soviet collapse, the aboriginal allocation had been reduced by 90 percent, and only 55 Amur aboriginal communities remained. At least 25,000 people were living, at least on paper, with no means of support whatsoever.

But fish and the communities that had long depended on them weren't the only victims of the brave new world in the Russian Far East. Kamchatka's huge brown bears were slaughtered for their gall bladders, for which superstitious Chinese businessmen would pay as much as $30 U.S. a gram in the hope of a cure for impotence. The Amur tiger—the world's largest—was hunted for the money that rich Chinese eccentrics were willing to pay for the tiger's bones, which were brewed into a broth that was said to contain the elixir of life.

And it wasn't just the Russian Far East that was being pushed to the brink of ecological collapse. In the most dramatic and rapid population crash biologists have ever observed among mammals, more than 90 percent of the world's saiga antelope were swept from the Russian steppes and the plains of Kazakhstan during the 1990s. When the decade began, more than a million saiga roamed the area, and for generations hunting had been sustainably managed. The saiga were so numerous that the World Wildlife Fund proposed that their horns could be marketed as an alternative to rhinoceros horns in the Chinese folk-medicine trade. By 2003, only 30,000 saiga were left. One giant herd in the Betpak-Dala region of Kazakhstan suffered a 99 percent drop in its numbers. The herd began the 1990s with more than half a million animals; a dozen years later, only about 4000 remained.

The vanishing of the Amur giants was actually part of something else that was going on throughout the temperate world, and in that larger phenomenon, there were the same nuances of meaning, the same problem of terms that required some elaboration. There were poachers, and then there were poachers. There were proximate causes and ultimate causes.

There were global forces at work that weren't accounted for by either of the ideologies separated by the great gulf the Cold War had cleaved upon the earth. Those forces are present in the very nature of complex human societies. Left unchecked, powerful societies tend to draw down the biological capital supporting distant peoples, in distant places. In Russia, it was as Misha had said, about how the tyranny of the tsars was still a dark shadow on the waters of the Amur River. But around the world, those same forces were operating in different ways.

Among all the earth's species, human beings are especially good at allowing their reach to expand beyond their immediate horizons. Humans are very good at breaking out of the ecosystems that limit the growth of local human populations and limit consumption of natural resources. The old "feedback loops" that tend to restrain over-harvesting of local renewable resources—technological limits, and market limits that link local survival to sustainable harvests of local resources—are breaking down, all over the world. To watch those forces at their busiest, especially in the temperate world, the best place to stand is on a beach, or on a dock, looking out across a river, or out to sea.

Before the twentieth century, the total yearly catch of all species of fish from the world's rivers, lakes, and oceans had never exceeded

10 million tonnes. In the first half of the twentieth century, the world's fishing fleets switched to steam engines, then engines powered by fossil fuels, and quickly developed the capacity to haul unprecedented volumes of biomass from the sea. After the conflagration the Russians call the Great Patriotic War, heavy winches and high-performance diesel engines allowed the world's fleets to push farther offshore, where they could catch even more fish. By 1950, the annual global catch had doubled, to 20 million tonnes. Then came refrigeration technology and even better engines, and the fleets pushed even farther offshore and deeper into the world's more remote coastal regions. The fuel subsidies that the "capitalist" countries of the industrialized world provided their fishing fleets made the Soviets look like pikers.

By the end of the 1950s, the global fish catch had doubled once more, to 40 million tonnes. And by the end of the 1970s, the global fish catch was growing faster than even the rapid rate of human population growth: it doubled yet again, to 80 million tonnes. Increasingly, the catch was coming from places beyond the reach of law, beyond the regulatory zones of nation-states. Globalization allowed Atlantic bluefin tuna to find its way to the Tokyo fish markets in one day, and global capital made it even easier for the great powers to ensure that the costs of overfishing were borne by other people. But by the 1980s, the oceans were refusing to give up any more fish. Around the world, huge fish stocks were collapsing.

It is hard to say just how much of this is resulting in the absolute extinction of fish species. The oceans are believed to contain roughly half of the vertebrate species on earth, but comparatively little is known about the diversity of life in the sea. The International Union for the Conservation of Nature (IUCN) reckons, for instance, that almost all the world's bird species are already known, and that of the roughly 10,000 known bird species, 1 in 8 is threatened with extinction. Among the world's mammals, which aren't believed to be anywhere near as well documented as birds, about 5400 species are known. Scientists affiliated with the IUCN had surveyed about 4800 mammal species by 2004, and found almost 1 in 4 threatened with extinction. Fish, however, are a thorny problem. A whopping 28,000 fish species have been described by scientists. That is an enormous inventory, but it is nonetheless considered a small sample of the number of species in the sea. Only 1700 of these had been surveyed by 2004, but almost half of that number are considered to be facing some threat of extinction.

During the early 1990s, the centuries-old cod fisheries of the North Atlantic collapsed almost overnight. But the boats kept fishing, because they could. When the big fish like tuna and cod became scarce, the boats started catching other fish, the ones the giants used to prey upon, small fish like sardines and anchoveta, and then the giants started to disappear. Fisheries scientist Daniel Pauly was the first to document the chain of events. Through the late twentieth century, global catches fell only slightly, but the big fish were being replaced by small fish. In the North Atlantic, the biomass of big fish declined by two-thirds during the second half of the twentieth century. Aquaculture production soared, but it wasn't relieving the pressure on wild stocks: it often takes four kilograms of small wild fish, which are caught and turned into feed pellets, to produce a single kilogram of big, farmed fish, like salmon. Still, the boats kept fishing.

By the first decade of the twenty-first century, the prestigious international journal *Nature* could confidently report that 90 percent of the giants of the watery part of the world simply weren't there anymore. All but gone were the big tunas, the marlins, the swordfish, the halibut, the sharks, and the skates. In each case, the time it took for each of the high-seas stocks of big fish to be reduced to roughly 10 percent of their former abundance was, on average, 15 years. It was roughly the same amount of time it took to almost completely empty the River of the Black Dragon of its taimen, its sturgeon, and its kaluga.

The *Nature* study was undertaken by Ransom Myers, of Dalhousie University, in Halifax, and Boris Worm, of the University of Kiel, in

Germany. Their findings were a shock, and initially many fisheries biologists had their doubts. But the data proved sound. According to Jeremy Jackson, of the Scripps Institute of Oceanography, the findings had been so hard to accept because everybody had forgotten what giants once lived in the seas. "We had oceans full of heroic fish," Jackson said, "literally sea monsters. People used to harpoon three-meter long swordfish in rowboats. Hemingway's *Old Man and the Sea* was for real."

With the giants gone, many fishermen were reduced to catching tiny shrimp, even jellyfish, and in the netherworld of fisheries-management regimes that had presided over all this, it was hard to draw clear lines between proximate causes and ultimate causes of the collapses, between what was poaching and what was lawful. Poaching wasn't happening only in the criminal state that Russia had become, or just in the so-called developing world. It was happening on the east coast of the United States, where Georges Bank cod were mined until their numbers were reduced to a fraction of 1 percent of their former abundance. On Canada's east coast, 99.9 percent of Newfoundland's northern stocks were hauled up out of the ocean. Up and down North America's west coast, rockfish, some more than a century old, were being dragged from the bottom of the sea in such numbers that no scientist could imagine their recovery. The fish were ancient mariners, as beautiful as their names. They were chillipeppers, auroras, and vermilions, duskies, redbands, and red Irish lords. In the Indian Ocean, and off the coasts of Senegal and Guinea Bissau, and it was happening by trawling and gillnetting and longlining.

The murkiest water lay between what was poaching and what wasn't. As I was writing this book, the world's largest fishing vessel was the *Atlantic Dawn,* a 144-metre ship powered by engines pulling 28,730 horsepower. The *Atlantic Dawn* fishes with a net that could engulf a football stadium. To evade European Union rules limiting the European fleet's catching power, the ship's owners simply registered the vessel as a merchant mariner and went fishing off the coast of Mauritania.

It was a much smaller vessel that pulled up on the beach one sunny afternoon at Sikachi Alyan, a Nanai village of about 300 people on the Amur River, about a hundred kilometres downstream from Khabarovsk. It was a shallow draft rowboat, built of rough boards. The man who had been rowing it hurried over and joined us.

I'd been walking along the willow-shrouded beach with Nina Ignatieva, the 48-year-old village chief. I'd been trying to explain that everything about the Lower Amur felt like the Bulkley Valley in northern British Columbia. The trees were smaller, and there were no mountains here except in the far distance, but the people looked like they could have been Athapaskans, which made some sense, since the Amur was a crucible of the peoples who eventually settled in North America. Sikachi Alyan was just like an old Bulkley Valley village, with little square-log houses strung out along a bluff above the river. Ignatieva had a fair command of English. She was fluent in Russian, and spoke Nanai and a bit of German, too. A small, sharp-featured woman, with thick black hair tied up in a bun and wrapped in a colourful silk scarf, she'd been a schoolteacher once. We were walking among Sikachi Alyan's riverbank stonehenge of petroglyph boulders, works of ancient art that had made the village famous in the Russian Far East. A group of us had come from the symposium to see them.

The man from the rowboat walked up the beach to us, smiled, and slapped me on the back. He was wearing a tattered grey windbreaker over a white shirt with two or three buttons left, and Ignatieva said the amulet that hung from his neck was the kind that warded off stomach ailments. The man smiled broadly, showing a few chipped teeth. We shook hands, and Ignatieva introduced us. He was another Misha, a 53-year-old father of four, with two grandchildren. "He is an unemployed person from the village," Ignatieva said. I thought this an odd way to describe him, since we'd just been talking about how nine of every ten adults in the village were unemployed. But calling Misha unemployed drew attention to the fact that he had just come from a small fishing boat, so I asked what he'd been doing. Ignatieva said that he had been checking whether any fish were caught on the hooks from the sunken set-lines he'd hidden out in the river.

Before the man from the rowboat arrived, Ignatieva had been talking about the old days, when people had jobs, Sikachi Alyan kept its own herds of cattle and horses, and her mother looked after the village pigs. Back then, fish were plentiful in the river, she said. "But after perestroika, everything fell apart, and it is different now," she said, "and everybody is dividing into rich people and into poor." For several years, the village had been allowed only a small subsistence fishery, mainly for chum salmon, so between salmon seasons the villagers fished illegally. "They catch the *bilaribitsa,* the whitefish," Ignatieva explained. "They get it from the river and they sell it to those who come here from Khabarovsk, because the fishermen, they have to poach. It is the single way to get some money, for bread and butter for their children."

We continued on among the petroglyphs. The earlier ones are believed to have been carved at least 5000 years ago, and the place had been an important centre for shamanic ritual. Nobody knows much about the old ways anymore, but the Nanai are still fiercely proud of the place. The petroglyphs came from the time when the River of the Black Dragon got its name, in the days before the Manchus branched off from the Nanai's ancestors. Almost every boulder was adorned with spiralled designs, anthropomorphic figures, and representations of birds and fish and moose, but it was getting harder to make them out, because the afternoon sun was slowly falling into the forests of the west.

The Nanai of Sikachi Alyan are only a shadow of the people the American adventurer Perry McDonagh Collins encountered while rowing down the Amur in 1857, and the Amur is no longer the salmon-rich river of Collins's journals. Collins was scouting a route for a round-the-world telegraph cable, a vision that came to an end when the Atlantic was spanned by a submarine cable in 1866. When the Amur River turned north for its final descent to the Sea of Okhotsk, Collins was routinely greeted by boisterous Nanai women paddling alone in canoes. Strong and beautiful, they wore elaborate salmon-skin tunics trimmed with brass coins and seashells. They were adorned with several-stranded earrings and white-metal nose rings. These "laughing, frolicking damsels" were as "unabashed as any well-bred lady of London or Moscow," and the men Collins met came in flotillas of canoes, bearing sable and black-bear furs to trade for silver. The country was like the tropics, Collins noted, "from the variety and the richness and the foliage of the trees."

It's too bad nobody remembers the stories about this place, Ignatieva said of the petroglyphs. Still, the fishermen and the hunters sometimes feed the spirits, she said. It is a simple ritual. "The spirits of the river, for example. Spirits of the forest. We put a little fire, and put some food in the fire and just feed the fire, and make a speech to the river. The fishermen and hunters, they do this." But the meaning of the petroglyphs and of the old rituals that went on here have been forgotten. When Ignatieva was growing up, nobody was allowed to come near the place. "It was taboo," she said.

There had been a time when the Soviets had printed books for the Nanai, in their own language, but after Stalin came to power the books were burned and the shamans were packed off to the gulags. Like their neighbours the Nivkh and the Ul'chi, the Nanai were, by the 1990s, in the midst of language extinction. Many of the older people spoke only Nanai, and the middle generation, people Ignatieva's age, were usually bilingual. The children playing along the beach near the petroglyphs that afternoon were the first Nanai people in history who could not speak their own language. They knew only Russian.

"This one, I think, is a tiger," Ignatieva said, running her small hands along the deep grooves of another petroglyph. The talk turned to the family of tigers that had lived on a nearby island in the river over the winter, and we came to another petroglyph. "Maybe this one is a bear," she said. "This one I think is a girl. Can you see it? Yes. I am sure." Misha from the rowboat pointed to another one, and spoke excitedly. "He thinks maybe it is a fish," Ignatieva explained. It did look like a fish, but Ignatieva said she had been told it was meant to be a deer.

I wanted to know how the day's fishing had been, and Ignatieva asked Misha for me. "He says there are no fish. He says he has no idea where they went." Ignatieva spoke sternly to Misha and then explained that she had told him it would be better for him to make something, to carve things, or make amulets that visitors might buy. I asked Misha why he was fishing. Ignatieva translated his reply: "Of course, the fishing is prohibited here. But how am I supposed to live? If I don't catch fish, I will not eat."

We walked on through the petroglyphs. "This is a mask," Ignatieva said. And then, at another boulder, "and this is a mask too." I could make out only what appeared to be concentric circles. "There is another like this. Would you like to see? Come this way."

We looked at some more petroglyphs, but the sun was now falling behind the willows. Misha was speaking again and he was gesturing as he spoke. I could see he wanted Ignatieva to tell me something. "He says there is supposed to be fish in the Amur River. He says he doesn't know where they went." Misha spoke again. "He says he will go fishing tonight," Ignatieva said.

"He says if he catches some fish, his children and his grandchildren will have something to eat tonight. If not, they will not eat."

A lion

The Ghost of the Woods

Today, the thing that stares me in the face every waking hour, like a grisly spectre with bloody fang and claw, is the extermination of species. To me, that is a horrible thing. It is wholesale murder, no less. It is capital crime, and a black disgrace to the races of civilized mankind.

—William Hornaday, head of the New York Zoological Society, 1913

On a perfectly ordinary late August afternoon in Port Alice, a pulp mill town on the northwest coast of Vancouver Island, David Parker decided to go for a walk. The retired mill maintenance foreman had been working on the roof of his small house when he felt a bit of a cramp in one of his legs. He reckoned a walk would do him good, so he set out on the route he was accustomed to taking on his evening strolls.

Parker headed out of town on a gravel road that connects Port Alice with the Jeune Landing log-sorting yard down at Neroutsos Inlet. Just over a kilometre from Jeune Landing it started to pour, and Parker ducked under a rock ledge at the side of the road to wait out the rain. He thought he heard a noise behind him. When he turned to see what it was, he found himself staring into the eyes of a cunning ambush predator that has gone by many names. It has been called night screamer, night crier, swamp devil, cougar, panther, catamount, mountain lion, and ghost walker.

Taxonomists call it *Puma concolor.* It is a survivor from a nightmarish bestiary of long-extinct carnivores that included *Titanus,* a flightless, horse-sized bird; *Arctodus,* a gigantic bear that was almost certainly the largest mammalian carnivore in the earth's history; and *Teratornis,* a cousin of the California condor the size of a small airplane.

The animal poised in a crouching position a few centimetres from Parker's face was a healthy young male. Parker turned to run, but the creature pounced, and Parker fell face down in a shallow ditch. The cougar was on Parker's back, sinking its teeth into his skull. Within seconds, most of Parker's scalp had been torn away, and his jaw and left cheekbone were broken. The orbital bones of his left temple had been crushed. His right ear hung by a thread of skin.

Parker found himself thinking, well, this is where everything ends. But then he decided he wasn't going to die on a rainy afternoon in a shallow ditch beside a dirt road on the outskirts of Port Alice. While the cougar was tearing at him with its claws and pulling away bits of his scalp, Parker reached for the small pocketknife he always kept in a tiny sheath attached to his belt. You fight back, he said later. You fight back, because that's all there is left to do. As Parker reached for his knife, the cougar bit his face, and his right eye protruded from its socket. Still, he managed to open the eight-centimetre folding blade, and he plunged it into the cougar's neck. With his other hand, he held onto the animal. Moments later, the cougar stopped struggling and gave up its final breath. With blood streaming from his face and head, Parker stood up and slowly started walking back to Jeune Landing. "I just put one foot in front of the other," he remembered.

"I just thought I'd try to get as far as I could before I collapsed. That's all I had on my mind."

At Jeune Landing, a millworker found him staggering down the road. Parker was flown by air ambulance to the Royal Jubilee Hospital in Victoria, at the southern tip of Vancouver Island, where he spent more than ten hours in surgery. Over the next two years, Parker would make more than 30 return visits to doctors and surgeons at the Victoria hospital, a seven-hour drive south. It took 350 staples to secure his scalp to his skull, 200 stitches to close the gashes on his face, and several plates and metal screws to reconstruct his jaw. But despite the constant pain, recurring nightmares, and anger, eventually Parker became stoic about it all. He got back to playing old-timers' hockey, twice a week. "I was in the wrong place at the wrong time," he'd say. "That's about all there was to it."

In Port Alice, a town of about a thousand people nestled against the rainforest south of remote Quatsino Sound, cougar attacks weren't exactly unheard of. Around Port Alice house cats routinely fell prey to them. In the towns and villages around Quatsino Sound, Winter Harbour, Coal Harbour, and Holberg, it was the same. Cougars came with the country. Vancouver Island was the cougar's last major redoubt in North America. Vancouver Island is the largest island on North America's west coast. It's about a third the size of Ireland, and its lush temperate rainforests were teeming with coastal black-tailed deer. There were no grizzly bears or lynx on the island. It all added up to perfect cougar habitat. Still, something strange was going on. In the three years before Parker was attacked in 2002, three cougars were shot within sight of his house. One was killed after it was seen stalking children on their way to school; it had been spotted in Parker's driveway. Another had attacked a dog on a trail behind Parker's place. The third was shot after becoming a routine nuisance at Port Alice's public works yard. The year before Parker was attacked, Eliot Cole was driving home from his shift at the pulp mill when he found a cougar on top of a man in the middle of the road just outside town. John Nostdal, a Seattle tugboat captain, had been riding his bicycle back to his anchored tugboat from Port Alice when a cougar jumped him from behind, knocking him off his bike. Cole had to beat the cougar about the head, first with his lunchpail and then with Nostdal's bicycle, to get it off the man.

In the ten years leading up to Parker's grisly encounter on the road to Jeune Landing, a cougar had walked onto the playground of the elementary school at Kyuquot, about 50 kilometres across the mountains from Port Alice, and killed an eight-year-old boy. Just south of Kyuquot, at Gold River, a seven-year-old boy was walking to school when he was attacked by a cougar and dragged into some bushes. RCMP constable Rick McKerracher shot the animal, saving the boy's life. A few months later, McKerracher, out riding his horse, was himself attacked by a cougar. Then a logger was attacked over in Zeballos. Then two campers were attacked south of Port Hardy.

The attacks could have been dismissed as an especially nasty spate of incidents in a small part of the world known for its cougars, except that it wasn't happening just on Vancouver Island. Cougars were attacking people everywhere.

In 1991, University of Arizona wildlife ecologist Paul Beier published a detailed study of cougar attacks in North America between 1890 and 1990. He had searched records in newspaper files, popular magazines, and academic journals. He'd contacted wildlife agencies in all 12 western U.S. states, as well as in Alberta and British Columbia. What Beier found was that throughout North America, more people had been attacked by cougars since 1970 than during the entire 80-year stretch between 1890 and 1970. Of 53 confirmed attacks between 1890 and 1990, 36 had occurred after 1970. More people were being killed by cougars, too. Only four fatal attacks were reported between 1890 and 1970. Between 1970 and 1990, there were five.

In 2001, Beier revisited the subject and updated his 1991 survey, and these later findings were even more disturbing. Of the 98 cougar attacks in North America between 1890 and 2001, almost half had occurred during the 1990s. Between 1990 and 2000, another seven people had been killed.

Among the dead was an 18-year-old high school athlete who went for a run on the outskirts of Denver, Colorado, in 1991 and never came back. Another jogger, a young woman killed by a cougar near Sacramento, California, in 1994, became the first person killed by a cougar in California since 1909. In a widely reported and particularly tragic event in 1996, Cindy Parolin died while protecting her children from a cougar near the town of Tulameen in British Columbia's southern interior. In January 2001, Frances Frost, a 30-year-old resident of Canmore, Alberta, was killed while cross-country skiing in Banff National Park. It was the first death attributed to a cougar in the park's history.

Cougar sightings were also on the rise, and not just in those remote parts of Canada and the United States where the animals had persisted in relative abundance. They were being reported with increasing frequency in places where everybody had thought cougars had been hunted out generations before.

It was as though a particularly frightening creature had stepped out of the pages of a book about extinct ice-age megafauna and was walking among us again. And that's more or less what was going on.

Just how this could have happened is part of a story that defies closely cherished ideas about pre-industrial "nature" as a static and pristine state, separate and apart from human culture. It is a story in which people were already transforming the planet's ecosystems 30,000 years ago. In that story, North America was not peopled by the shy hunter-gatherers of conventional environmentalist narratives, the timid and cherubic "noble savages" imagined by such early romanticists as Thomas Burnet, William Gilpin, and Jean Jacques Rousseau. By the time Europeans arrived in

North America, most of the large mammals that the continent's first aboriginal peoples had encountered were extinct. The "wilderness" the Europeans found had already been radically transformed by people. The ghost walker, *Puma concolor,* is an evolutionary witness to events that were never supposed to have happened.

A good place to start the story is back in the Russian Far East, where we were wandering along the Amur River in the last chapter.

It was from the Russian Far East that the peopling of the New World began. In its conventional version, the story begins with the descendants of a handful of Asiatic tribes from the headwaters of the Lena River, just across the Stanovoy Mountains from the headwaters of the Amur. A small group of ice-age hunters with a voracious appetite for mammoths and mastodons ended up in North America by crossing the now-submerged subcontinent of Beringia sometime around the close of the long Pleistocene epoch—the period generally known as the ice age. About 11,000 years ago, they found a way south out of Alaska, through a long and frigid corridor that had briefly opened up between the Laurentian and Cordilleran ice sheets during a break between one of the many on-again, off-again ice-age spasms. By the time they stumbled out upon the vast and unpeopled heart of the North American continent, they were highly proficient in the pursuit of big-game animals and masters of an elegant, specialized toolkit that included spearpoints of a type first discovered at Clovis, New Mexico, in 1932.

It's because of that discovery in New Mexico that those people came to be called Clovis people. In archeology textbooks going back to the 1940s, the Clovis people and their kin ended up settling just about every New World nook and cranny, from the arctic tundra to Tierra del Fuego, at South America's southern tip. Because "Clovis" points and their associated artifact types are never found to be more than about 11,000 years old, it has always followed that the Clovis people are the ancestors of all "Indians" and that they arrived in North America little more than 11,000 years ago. It is known that this was a time of rapid and dramatic climate change, when the earth was finally warming after the long Pleistocene winter.

What is also certain is that throughout North and South America, this also was a time of mass extinction. At the onset of the Holocene—the epoch in which we are now living—four-fifths of all large vertebrates in North America suddenly disappeared. Conventionally, the die-off has been blamed on the broad-scale ecological change underway at the time. But conventional theories about those extinctions, and about the part people might have played in them, are undergoing some serious second thought, for a whole host of reasons.

The main reason the old version of events stopped making much sense was that a kind of safety zone had always separated almost all the megafaunal extinctions of the early Holocene from the arrival of people in North America, and that safety zone was vanishing. This was largely a consequence of archeological discoveries that pushed the date of humans' arrival in North America back before 11,000 years ago. It was also because of advances in scientific understanding about such things as the shape and extent of the Pleistocene ice sheets, which we now know did not present as formidable a barrier to people as it seemed when the "Clovis-first" theory was developed.

A central character in the new story science has begun to tell about those early days is a little hunter-gatherer known to us only as Arlington Woman. She lived on Santa Rosa Island, off the California coast, and she appears to have had a bad fall while chasing pygmy mammoths with a pointed stick a long, long time ago. Her thighbone was found sticking out of a grassy ledge at the top of a canyon, about a kilometre from the sea, in 1959. Whatever happened on that fateful early Holocene afternoon, Arlington Woman never got up again.

After the initial 1959 find, nobody took another close look at the evidence from Santa Rosa Island until the 1990s. That second look at

Arlington Woman's thighbone, and at the detritus of organic material in which it was encased, threw a spanner into the works of conventional thinking about North America's human history and about the continent's early Holocene species extinctions. Arlington Woman now appears to have been living on Santa Rosa Island as far back as 14,000 years ago, which would have been long before any Clovis people were around. She was also evidently a member of a fully adapted local culture of islanders, a community well established by the time the Clovis people stumbled out of that horrible wind tunnel between the continental ice sheets. The Santa Rosa people were a maritime people, living several miles from the mainland. They were seafarers.

What this implies is that whatever might be said about the origins and the wanderings of the Clovis people, something else was happening, involving people who were different from the Clovis people's ancestors. The story of these other people has come to be called the coastal-route hypothesis.

That story likely begins on the lower reaches of the Amur River, in the general vicinity of Sikachi Alyan, the village where the meaning of ancient petroglyphs has been forgotten. The people here made a different journey, one that didn't involve gruelling treks across the Beringian tundra. Instead, they set out on a series of relatively short trips by boat, around the top of the North Pacific. The journey in this version of events got underway in earnest near the Kamchatka peninsula, on Russia's Pacific coast, and it involved a lot of island-hopping. The people proceeded along Beringia's southern shore to islands in the Gulf of Alaska that were never covered by the great glaciers of the Pleistocene epoch. Then they made several extended stays on a series of now-submerged islands off North America's west coast.

The coastal-route theorists have a good deal of support in the linguistic and genetic evidence, and by the late 1990s they could also boast a collection of artifacts found on the muddy sea floor just north of Port Alice on Vancouver Island, where David Parker was jumped by a cougar. The Canadian hydrographic survey ship *Vector* dredged up the artifacts just off the archipelago known alternatively as Haida Gwaii and the Queen Charlotte Islands. The artifacts aren't much to look at. They're mainly crude bits of stone from which spearpoints and whatnot have been worked. But they were found on parts of the seabottom where there were once river bottoms and estuaries of a coastline that was perfectly capable of supporting human communities, at precisely the same time that the continent, in the Clovis-first version of the story, was supposed to have been ice-bound and completely isolated from the rest of the world.

All this evidence puts human beings in the New World at least 2000 years earlier than previously thought, among an astonishing assortment of animals that didn't know enough to be afraid of humans. There were lions, almost exactly like African lions or the lions of the Gir forests of India, only about twice as big. There were armadillos the size of Volkswagens, and beavers as big as bears.

Those animals were there. People arrived, then the animals were gone.

Even the most devout Clovis-firsters don't deny that at least three dozen large-mammal species started disappearing from North and South America about 13,000 years ago, and by about 9000 years ago none were left. There were sabre-toothed tigers and scimitar cats and stocky, long-fanged dire wolves, and then there weren't. The stag moose, a swamp-dwelling giant with massive antlers shaped like the palm of a human hand, disappeared with the horses, camels, and four species of ground sloths—slow-moving, big-footed vegetarians that looked like a cross between an ape and a huge dog. One giant sloth that ranged from Texas to the Argentinian pampas was bigger than an elephant. Recent evidence suggests it may have been a carnivore.

It is probably no coincidence that one of the few animals from the old order that survived the early Holocene extinctions in North America is the fastest-running herbivore on the planet, the pronghorn antelope. Once massing in herds that rivalled those of the buffalo, pronghorns can reach speeds of 90 kilometres an hour and are second only to cheetahs in world animal speed records. But if you can't run, you might want to hide. That was the strategy of the haplodon, a pre-Pleistocene relic that has persevered into the twenty-first century, in pockets of several isolated subspecies in the U.S. Northwest and British Columbia. It spends its life almost entirely in underground tunnels, emerging only for brief nighttime forays, to gather shrubs, nettles, grasses, and ferns. The Columbia tribes knew the creature as the *sewellel,* but it is still so poorly known to scientists that well into the twentieth century the standard reference literature on *Aplodontia rufa* persisted in claiming that it was capable of emitting loud booming sounds and wept great tears when distressed.

For many of these extinctions, the arrival of human beings may have unhappily coincided with the creature's demise from climate-change effects. Multiple extinctions could have resulted from rapid ecosystem change set in motion when humans exterminated one or a few large ecosystem-shaping animals, such as mammoths and mastodons. Some extinctions may have at least partly resulted from the sudden arrival of competing or predatory species that accompanied people in their southward journey around the ice sheets that isolated the arctic parts of North America from everywhere else. Others may simply have been the unhappy result of a combination of things. But the fossil evidence contains nothing to suggest that North America's megafauna were suffering any climate-related deprivation or ill health before they disappeared, and the archeological record is full of sites where stone tools are scattered among the bones of mammoth and mastodons, ground sloths, stag moose, peccaries, and giant beaver.

When human beings moved into Australia and New Zealand, scattered throughout Polynesia, and sailed to Madagascar, the islands of the Mediterranean, and the islands of the Caribbean, the pattern that followed was the same. Humans showed up and animals started going extinct, one after another. In Australia and New Guinea, an extinction spasm carried away at least two dozen species of large vertebrates not long after people arrived. The first Australians, who landed about 50,000 years ago, discovered an island continent fabulously rich in marsupial species that are now extinct. Diprotodon, a distant relation of modern-day wombats, stood two metres at the shoulder and looked a bit like a bear but appears to have been exclusively herbivorous. The largest marsupial ever, Diprotodon survived in Australia until about 40,000 years ago. Its disappearance coincided with the vanishing of a marsupial lion that was almost certainly its major predator. Around the same time, the 200-kilogram short-faced kangaroo disappeared. It was a weird-looking kangaroo, with grappling-hook claws and hooves like a horse. The Tasmanian wolf was also a marsupial, and while it is well known for its disappearance from the Tasmanian wilds in the late nineteenth century, it once roamed Australia too. It was gone from that island continent, and from New Guinea, by about 2000 years ago.

Until roughly 40,000 years ago, Genyornis was a flightless bird that stood as tall as a man. It vanished at the same time as the bear-sized wombat, the marsupial lion, the short-faced kangaroo, and all the rest, and its bones are commonly found with artifacts from ancient campsites. Whatever it was that killed all those creatures, it was an event that carried away at least 19 large Australian marsupials. More than 80 percent of all Australian species weighing more than 150 kilograms were lost. The toll included a cow-sized, meat-eating goanna, a distant cousin of the Komodo lizard of Indonesia. Also gone was the quinkana, a fast-running land crocodile that grew to lengths of seven metres.

In 1999, a team of researchers led by geochronologist Gifford Miller, of the University of Colorado, found that the fires Australian aborigines set to improve plant productivity may have been the tipping point that doomed many species. Sediment cores show a sharp spike in charcoal content around the time humans arrived in Australia. Over time, if enough shrubs were burned, the populations of browsing herbivores would decline to an extent that they would become gravely vulnerable to predation. Soon enough, they'd be gone. And soon enough, they were.

Similarly, on the islands of New Zealand, the only great cataclysm that can account for the extinction of dozens of bird species is the arrival of Polynesians, with their rats and dogs, about a thousand years ago. The most peculiar thing about New Zealand is that birds played the same ecological role as mammals elsewhere in the world. The islands were home to 11 known species of moa, all wingless birds, from the small 20-kilogram *Euryapteryx curtus* to the giant *Dinornis giganteus,* which stood at least two metres tall. Within the first five centuries of Polynesian settlement, they were extinct. Another two dozen bird species also vanished around the same time, including several flightless geese, ducks, and rails.

Extinctions followed the Polynesians wherever they went. As many as 2000 bird species are believed to have become extinct on the tropical islands of Oceania in the years following human settlement. In the Hawaiian islands, which Polynesians first reached in their massive sea-going catamarans and outriggers about 1700 years ago, the non-human settlers did at least as much damage as the people. Among the passengers on those first Polynesian vessels were chickens, small-bodied pigs, dogs, and an especially nasty creature, the tropical rat, almost certainly a stowaway. Fire, land-clearing for taro fields and sweet potato, hunting, and the carnage wreaked by introduced species resulted in the disappearance of at least 50 Hawaiian bird species, long before Europeans showed up. A similar event occurred on the islands of New Caledonia, 900 sea miles east of Australia. The bones of at least a dozen extinct bird species have been found in cave deposits left about 3000 years ago, when human beings arrived.

Madagascar's extinctions follow the same trajectory. This island off the east coast of southern Africa lost several species of giant lemur, giant tortoises, an "elephant bird," and a pygmy hippopotamus. The extinctions followed quickly upon the arrival of mariners from Indonesia about 2000 years ago.

It isn't much of a mystery why so many species started disappearing about 5000 years ago from the islands of the Caribbean. No broad-scale ecological change was underway, certainly not of the same magnitude as the recession of the Pleistocene. The only thing happening back then was people were arriving. They showed up in the Caribbean roughly 7000 years ago. When they came, sloths still lived in the forests, including the Cuban ground sloth, which grew to the size of a black bear. The last of those species vanished about 3000 years ago. The hutia, a plump and furry animal, a bit like a huge guinea pig, grew to 15 kilograms. Its bones have been found in Puerto Rico, Haiti, the Dominican Republic, and the Virgin Islands, often in the "kitchen midden" refuse heaps of aboriginal villages. Hutias were among an unknown number of species of at least ten types of rodent genera that disappeared from the islands over the years. They are believed to have held out in the more remote parts of the islands until Europeans arrived in the late fifteenth century, but they're gone now. Also gone is the Saint Croix macaw, known only from bones in middens.

Early Europeans wreaked their own carnage, and the story of Eurasian extinctions reveals some of the saddest surprises. For one, there really was a unicorn. At least it's hard to describe the creature any other way. Elasmotherium was a huge and shaggy animal, a distant cousin of the woolly rhinoceros. A huge horn that may have reached lengths of two metres grew straight out of its forehead. It is unclear whether human beings had any hand in its extinction, but the creature coexisted with us. Elasmotherium is generally believed to have vanished only about 10,000 years ago, but the oral traditions of the Evenks people of Siberia suggest that it may have survived in pockets. Chinese and Persian accounts suggest the same, and the tenth-century Arab scholar Ahmad ibn Fadlan, the Baghdad caliph's ambassador to the Volga River Bulgars, wrote of an animal matching Elasmotherium's description that was hunted with poisoned arrows in the Volga forests. The ambassador's account is likely quite fanciful, but it isn't all that far-fetched. A dwarf subspecies of woolly mammoth lived until 4000 years ago on Wrangel Island, off Siberia, and European bears have persisted into this century in such redoubts as the Carpathian Mountains.

Europe had its bison, beavers, and giant elk, and until Roman times even a remnant population of European lions. Herds of Pyrenean ibex roamed the mountainous border regions between France and Spain until well into the twentieth century. The last one, a 13-year-old female, was found dead on January 6, 2000, in Spain's Ordesa National Park. A tree had fallen on her, crushing her skull.

Of all the Old World's lost bestiaries, the islands of the Mediterranean were home to one of the strangest. Among the harder-to-believe animals was a giant flightless owl that stalked the forests of Crete. Pygmy hippopotamuses and miniature cave goats lived on several islands, along with the Prolagus, a sort of a cross between a rabbit and a guinea pig. On Sicily there was an elephant that at its largest was never bigger than a pony. All told, 30 large vertebrates disappeared from those islands.

Out of the sad story of all these extinctions, and all that carnage, some things should be fairly obvious.

The first is that whenever human beings traverse a new horizon, order collapses. Old feedback loops break down. We saw in Singapore that the epicentre of the current extinction crisis is in the tropics, where rapid technological change has whipped up its grisliest whirlwind in the industrial deforestation of tropical forests. The technology of high-performance diesel engines and heavy winches opened high-seas frontiers for the world's fishing fleets, quadrupling the amount of fish being hauled from the seas between the 1950s and the century's end. Order collapses and extinctions follow from cultural upheaval as well, whether it's the dying Nanai language in the village of Sikachi Alyan or all those endangered species for sale at the Khabarovsk fish market. Thousands of years earlier, the first people arrived in North America, Australia, New Zealand, and all those other unpeopled places, and order collapsed then, too. Mass extinction followed.

The second is that, except for the most isolated cases, human-caused extinctions do not occur because of wilful rapacity. There's just no evidence for it. In the long and sad story of human involvement in the extinction of animals, a "two legs bad, four legs good" story simply won't withstand scrutiny.

Remember that most of the bird species lost to extinction over the past 400 years vanished not because they were deliberately exterminated, but because of habitat loss and the depredations of introduced species, such as rats and pigs. This was the case even with the famously extinct dodo.

Some of the same unwitting accomplices in extinction—rats and dogs—followed Polynesians to New Zealand a thousand years ago, just before all those birds vanished. In the same way, the mass extinction that followed human beings throughout the New World of North and South America may well have been as much the fault of "new" animals that were skirting the fringes of the Pleistocene ice sheets around the same time.

Ignorance, greed, and an utter disregard for our fellow species are part of the story of "modern" people. There is no reason to doubt that the pace of deforestation, which accelerated during the twentieth century, is resulting in a scale of species loss that is completely off the charts. But just as it would be an unsupportable and vulgar assertion to attribute pre-industrial animal extinctions to the wilful rapacity of hunter-gatherers or any "aboriginal" societies, the notion that "modern" civilization has heedlessly and deliberately driven animals over extinction's cliff edge is also demonstrably wrong.

Fewer than a quarter of all known animal extinctions since 1600 can be attributed to hunting or deliberate extermination, and, as we have seen, "modern" people routinely make great sacrifices in aid of their endangered fellow species. About 40 percent of all that have disappeared since 1600 were the victims of introduced species, such as those rats from ships. Almost as many species, at least 36 percent, became extinct because of habitat loss or, perhaps more importantly, habitat fragmentation.

This is where it would be useful to remember another thing we faced up to back in Singapore: people change "nature," and nature forces changes back upon us and upon the way we imagine it. That's how the dialectic works. When we see the North American "wilderness," we're still looking at the great "concatenation of differently functioning and variously labelled mirrors" that E.E. Cummings wrote about.

Even when the great John Muir looked out upon his beloved Yosemite mountains, he was looking into a kind of mirror. One of the boldest Americans of his generation, Muir was a visionary, a vigorous public intellectual, and arguably one of the greatest spiritual leaders in American history. He was among the founders of the Sierra Club in 1892, and he was one of the first conservationists to fully appreciate the connectedness of all things. But Muir saw in the Yosemite a pristine and "wild" place, and if aboriginal people had played any role in its ecology, it was no more than that of the "birds and squirrels."

Muir did not see that the mosaic of landscapes in the Yosemite was in fact a consequence of ecological modification by the Miwok people, who had been burning, pruning, and harvesting there for several centuries. In the Rocky Mountains, moose were a rare sight before the late nineteenth century, and so were elk, those majestic animals that have become such a symbol of the mountain wilderness of the North American west. In 1995, Charles Kay, a scientist with Utah State

University, concluded that aboriginal communities in the Rockies managed berry crop production by routinely setting fires, so much so that they significantly altered the region's ecosystems, and by using fire and hunting suppressed elk populations to extremely low levels. "Based on historical and archeological data," Kay concluded, "there are now more elk in the West than at any time in the last 10,000 years."

It would be hard to understate the significance of the Edenic hunter-gatherer myth in the development of contemporary environmentalism. But regardless of its appeal, the berry-picking neolithic ecologist-philosopher stereotype fits many North American aboriginal peoples quite poorly. At the very least, the characterization of aboriginal peoples as innately spiritual hunter-gatherers living in some sort of sacred balance within stable ecosystems, and whose economies were simple bow-and-arrow affairs, wholly ignores the histories and the immense cultural and national diversities of the New World's peoples. Fondly held mythologies associated with the North American "wilderness" leave little room for contrary evidence about what the continent was really like before Europeans arrived.

It's true that the aboriginal peoples Europeans first encountered in North and South America were often living in a kind of harmony with their fellow species. Compared with the wilderness-hating Puritans and with the American expansionists forever pouring westward from the United States' 13 founding colonies, it wouldn't be hard to conform with the ideal of a stable-state, low-impact way of life, integrated within a flourishing ecosystem. A case can also easily be made for a kind of equilibrium at work between many nineteenth-century aboriginal societies and the prey species that supported them. This may have been especially so among the "horse cultures" of the Great Plains that developed during the eighteenth century, and of the ancient salmon societies of the Northwest. It is also true that many of North America's aboriginal societies had developed intricate and complex systems of natural-resource management.

But it's quite likely that after the initial wave of Early Holocene extinctions, most of North America's "Indians" gradually became town-dwelling agriculturalists, living within radically reshaped ecosystems. This is not a widely told story. The world is well aware of the Great Pyramid of Giza, for instance, and we're all happy to accept that ancient Egyptians accomplished something that can be described, without controversy, as civilization. But far fewer people have even heard of the pyramid at Cahokia, in what is now Illinois, just a few kilometres east of St. Louis, Missouri. The archeology there doesn't fit so neatly into what Europeans wanted to imagine North America was like before Christopher Columbus sailed across the Atlantic in 1492. The Cahokia pyramid, the central feature of a small city of about 20,000 people, was larger than any of Egypt's pyramids. It remained the largest structure ever built in the United States until the late twentieth century.

By the time the Spanish encountered it in the sixteenth century, the Inca civilization of the Peruvian Andes had already reshaped more than 600,000 hectares of mountains into terraces of small, carefully tended garden plots. Back on Canada's west coast, the first Europeans to look closely at the southern tip of Vancouver Island found it so altered by human hands that it resembled the pastoral English countryside. Before its fall to Hernando Cortés in 1519, the Aztec capital at Tenochtitlán was larger than Paris was at the time.

Some archeologists see wholly transformed landscapes even in those places long regarded as the darkest recesses of the world's wilderness. In the upper reaches of the Amazon basin, they see nearly eight million hectares of land raised above the flood plain by an elaborate network of berms and causeways constructed around the time of Christ. At about the same time, Hohokam people were excavating irrigation canals that would eventually run more than 2700 kilometres though the Sonoran Desert, in present-day Arizona, and aboriginal people in California were establishing artificial oases in the desert to support plantations of fan palms.

Just as many animals had vanished from the New World before anybody really got a chance to know them, so the vast majority of the western hemisphere's aboriginal societies were mere shadows of their former selves by the time they became known to the rest of the world. In much of the literature about these peoples, the depopulation effect of European epidemic diseases is still only barely acknowledged. Smallpox, influenza, typhus, diphtheria, whooping cough, bubonic plague, mumps, measles, and other diseases drove uncounted cultures into extinction. North America's human population shrank by as much as 95 percent. The continent's first peoples knew no immunity from or defence against these diseases, and so they succumbed.

Some of the continent's first European explorers were lucky enough to get a glimpse of what it was like before the diseases came. In the early 1540s, Hernando de Soto found dozens of small cities dotting a thickly populated landscape throughout what are now the U.S. southern states. A century later, the Coosa, Mississippian, and Caddoan civilizations he encountered were gone.

New England appears to have been at least as densely populated as the South and Midwest, but diseases had swept through the Atlantic coast forests, probably several times, even before the *Mayflower* pilgrims showed up in 1620. While contemporary Americans tell their children a heartwarming story of Wampanoags hosting European settlers at the first Thanksgiving with turkeys and yams and other fruits of the land, the reality is that the Plymouth colonists survived their first winter by robbing the corn caches of towns where the bodies of recently dead aboriginal people were lying in heaps, victims of a disease they had picked up from French explorers a few years before. In 1792, when George Vancouver traversed Juan de Fuca Strait, which marks the Canada–U.S. border on North America's west coast, he found huge but empty villages, and the faces of the few people he encountered in canoes were often horribly scarred from smallpox.

Estimates still vary wildly and academic debates still rage, but there is little evidence disputing the proposition that North America had been nearly emptied of its original inhabitants long before European settlers built their first major towns. The former orthodoxies about North America's pre-Columbian aboriginal population are slowly being abandoned. Those orthodoxies are epitomized by the Smithsonian Institution's 1928 estimate that put no more than 1.1 million aboriginal people north of the Valley of Mexico at the close of the fifteenth century. Later estimates by mainstream demographers and anthropologists have put North America's aboriginal population at 40 times that size. Some argue that the human population of North and South America, before the ravages of disease, may have been as high as 100 million.

Whatever the case, it now appears certain that long before Europeans came, perhaps for several thousands of years, human beings had been busy transforming North America's ecosystems. People caused the extinction of an unknown number of species, and people played some role in the emergence of wholly reshaped ecosystems that featured "new" animals—animals that filled all the ecological niches that had been occupied, before people came along, by North America's indigenous horses, camels, sloths, mammoths, mastodons, and the rest.

Among the new animals in the subarctic regions of the New World were grizzly bears, moose, mule deer, elk, and grey wolves. Also new were the buffalo that swarmed North America's plains in such great herds when Europeans first saw them. These are precisely the creatures we identify with North America's untrammelled and pristine nature, but we can just as easily make the case that most of these animals are functions of human intervention. In some cases, they really *are* new animals.

There's little evidence that the buffalo had roamed the middle of the continent for long in the kinds of numbers the first Europeans marvelled at. The fossil record suggests that the buffalo of North

America's prairies were actually very recent evolutionary descendants of a much larger, long-horned, more solitary Alaskan species that had spread south only a few thousand years earlier. The eastern subspecies of buffalo crossed the Mississippi from the west only about a thousand years ago. The buffalo the first Europeans encountered in Pennsylvania, for instance, appear to have arrived in a significantly human-altered habitat of open woodland only shortly before the Europeans themselves got there. Evolutionary biologist Tim Flannery goes so far as to describe the buffalo as a "human artefact," a creature that has evolved in response to human behaviour, mainly hunting pressure and fire. The immense numbers in which early Europeans first saw them may well have been an anomalous superabundance that occurred in response to the rapid, disease-induced decline in the numbers of their key predator—human beings.

The same may be said of those sky-darkening flocks of passenger pigeons, whose sad story has become emblematic to the standard narrative of human-caused extinctions. Flock sizes would have fluctuated over time, depending on climate, food abundance, and other factors, but those observed in the early nineteenth century were almost unbelievably immense. North America's passenger pigeons may have numbered five billion at their zenith. John James Audubon once witnessed a flock that darkened the skies from horizon to horizon and took three days to pass overhead. The sound was a roar, like a tornado, he noted. This must have been absolutely exhilarating to witness, but it seems unlikely that such enormous flocks would have been common in the days before North America was depopulated of its aboriginal peoples by European diseases.

Another symbol of the North American wilderness that appears to be a recent evolutionary innovation is the mule deer, the most abundant deer species in the western half of the continent. The mule deer is believed to be a hybridized descendant of white-tailed deer and a Pleistocene-era species of black-tailed deer. The mule deer is also one of the main prey species of that creature known to scientists as *Puma concolor,* the animal that suddenly started attacking and killing people all over North America in the late twentieth century.

Among terrestrial mammals, the cougar is second only to human beings in the range of New World ecosystems it has occupied. Just like humans, cougars have prowled the Yukon Territory in the north, and they've stalked game as far south as the shores of the Straits of Magellan. But along with the North American cheetahs, the sabre-toothed tigers, and the scimitar cats, cougars vanished from North America shortly after the first people showed up. They disappeared along with the dire wolves, the bear-sized beavers, the stag moose, and the giant armadillos.

Then they came back.

Several populations of cougars had persisted south of the Isthmus of Darien, otherwise known as Panama. They had been down there for at least two million years, having arrived from the north around the same time that Costa Rica's great diversity of bird species was taking shape in the mountainous middle sections of the land bridge forming between North and South America. From a 200,000-year-old lineage deriving from a locus of several cougar subspecies in South America, a new founder population of cougars moved gradually north, back into North America, recolonizing its long-lost cousins' old hunting territories. Into that vastly transformed continent that had taken shape through the centuries after the first people arrived, the founders quietly moved. Over time, a new cougar subspecies emerged. It was one of their number that attacked David Parker on the outskirts of Port Alice that August afternoon, on the road to Jeune Landing.

This much of the cougar's ancestry is clear from an astonishing bit of genetic sleuthing accomplished by four scientists, led by the University of Arizona's Melanie Culver, in the late 1990s. Culver and her colleagues analyzed genomic DNA material from 315 cougar specimens, including 54 from museums. The evidence of the cougar's disappearance and reappearance in North America is pretty solid, but what it doesn't show is just what caused the old cougars to disappear and what allowed the new cougars to flourish. The only event that can account for it, since cougars flourish in just about every sort of climate, is the disappearance of that nightmarish bestiary of Pleistocene species and the emergence of the "new" Holocene species that took its place. The new cougars seemed perfectly capable of surviving everything that the Clovis people and everybody else might have thrown at them. The mule deer were everywhere, and the cougars lived wherever people lived. For thousands of years, cougars flourished again throughout North America.

Then the Europeans came, and, again, order collapsed.

From the earliest days of European settlement in North America, cougars were hunted ruthlessly. In 1694, in Connecticut, a bounty of 20 shillings was offered for every dead "catamount," and in 1695, South Carolina required every surviving aboriginal person in the colony to kill a wolf, a cougar, a bear, or two "wildcats" every year, under threat of being subjected to a public whipping. Massachusetts was offering cougar bounties at the astonishing sum of 40 shillings in 1742, and "ring hunts" were common throughout the 13 colonies. An especially gruesome ring hunt was organized by a frontiersman with the unfortunate name of Black Jack Schwartz in Snyder County, Pennsylvania, in 1760. More than 200 hunters with dogs formed a circle about 50 kilometres in diameter, forcing all the wildlife toward a central slaughtering ground. The toll included 109 wolves, 112 foxes, 41 "panthers," 17 black bears, 98 deer, 111 buffalo, and at least 500 small animals. It is said that Schwartz, for his sins, was later ambushed, mutilated, and murdered by local aboriginal people.

Pennsylvania's last ring hunt was held in 1849, but few cougars were left, and they were officially declared absent from the state in 1871. By then, with the exception of Florida, where a sad remnant population persisted through the twentieth century, cougars existed east of the Mississippi mainly in the form of rumour. They were gone from New Jersey by 1840, from New Hampshire by 1850, from Vermont and North Carolina by 1880, from New York and Virginia by 1900. A small group may have held on in Connecticut until 1935, but for the most part, cougars had been extirpated from the eastern half of the continent by the end of the nineteenth century.

It wasn't just cougars that the early American colonists wanted to see gone. Puritan leader Cotton Mather was no John Muir. What Mather saw in the wilderness was the "devil's territories." To him and his co-religionists, the New England forests were dark and dangerous places, infested with all manner of demons and hellish beasts. In the Puritan view, clearing the forests to build a new "city upon a hill" and banishing the forest's beasts and its aboriginal infidels ever westward were not just happy events that coincided with European colonization: these were the very reasons for it.

The early American colonists were merely at the vanguard of a protracted campaign of European conquest and colonization that would spread not just throughout North and South America, but through Australia, New Zealand, India, and much of Africa. Extinctions followed Europeans wherever they went. But what was strikingly and historically distinct about the Puritan colonists was the extent to which they were animated by a pathological hatred of "wildlife" and "wilderness."

The early American colonists shared nothing with their contemporaries among the Dutch and the Portuguese, whose mariners have been blamed, quite unfairly, for the fate that befell the dodo, the famous bird of Mauritius. The Puritans were not comparable to the French and English settlers in Canada, whose fur trade purposes required the persistence of stable aboriginal communities and abundant animals, especially fur-bearers. The Puritans were like no other European colonial society on earth. In their strange belief that there was some kind of virtue in extermination, the Puritans were distinct. They were an anomaly. The founders of what was to become the most powerful nation on earth were extremists, and not just by our standards today, but by the standards of their own time. They were religious fundamentalists, Protestant separatists who considered themselves duty-bound, by their own radical interpretation of the Bible, to wage a war of extermination against the wild things of their new-found country.

Puritan values dominated America for at least a century after their arrival in 1620, and while this hatred of "wild" things was not typical of later American society, it was no sad accident that cougars ended up banished from North America's eastern forests. To be fair, cougars are not the kind of animal that a person can make friends with easily. Of the many names the early American colonists gave the animal, "night screamer" was not a foolish exaggeration. A cougar does not roar. It screams, and the bloodcurdling sound is unlike anything that any other animal emits. It has been compared to the agonizing scream of a grown woman being murdered. Cougars are beautiful creatures, but they kill people. They prefer children.

Driven from the landscape east of the Mississippi, cougars were pursued by bounty hunters across the western half of North America, and the first half of the twentieth century was a carnage. The animals were never very numerous on the Canadian prairies or the American plains, but in the far west, the bounty killings were astonishing. California paid bounties on 12,452 cougars between 1907 and 1963. In its final 22 years, before the program ended in 1978, Arizona's bounty program subsidized the shooting of 5700 cougars. Between 1937 and 1964, the Alberta government regularly paid bounties for 40 to 50 cougars a year. Across British Columbia, more than 16,000 cougars were shot for the bounty the B.C. government offered between 1910 and 1957.

It shouldn't be too surprising that nearly half the cougar attacks documented in North America from 1890 to 1990 occurred in British Columbia. The province is about the size of California, Oregon, and

Washington combined. The B.C. government's own tracking of cougar attacks showed 63 incidents in the province during the twentieth century, with about half occurring on Vancouver Island. More than a quarter of the attacks occurred during the 1990s, half of those on Vancouver Island, in places like Port Alice.

It also makes sense that Vancouver Island would be the epicentre of cougar attacks in North America. There are no grizzly bears on Vancouver Island, no lynx or bobcat, and very few wolves. The island is more than 30,000 kilometres in size. It's rugged, mountainous, heavily wooded, and sparsely populated. Even at the beginning of the twenty-first century, fewer than 700,000 people lived there, mostly around Victoria, on the south coast. The island's fishing villages and sawmill towns tend to be situated immediately adjacent to vast forests, and old-growth cedar and hemlock forests have always provided ideal habitat for coastal black-tailed deer and Roosevelt elk, two of the cougar's favourite prey.

But by the 1970s, reports of cougars were coming from all over North America, and from the most unlikely places.

Over the years, cougar sightings—in Missouri, Alabama, New York, Ontario, and elsewhere—were generally dismissed by wildlife officials as the result of overactive imaginations, which, in most cases, they undoubtedly were. It was as though people couldn't bring themselves to believe the cougars were really gone, so they saw them even when they weren't there—like the cries of eskimo curlews, heard long after they had vanished from the skies. For decades, throughout the eastern half of North America, cougars occupied a place somewhere in the realm of cryptozoology, a place normally occupied by lake monsters and hairy forest giants. But the early 1990s brought a rash of reported cougar sightings, some from the extreme northeast of the cougars' former range, and some from perfectly credible sources.

The last confirmed presence of cougars in New Brunswick was in 1841, but in February 1992, two wildlife officers in New Brunswick came upon the tracks of what they suspected was a cougar, in the forest near the town of Juniper. They didn't find a cougar when they followed the tracks, but they did find some excrement, which they sent to the Canadian Museum of Nature in Ottawa for analysis. The museum's curator confirmed that hairs in the droppings were from the hind legs of a cougar. Then, in May of that year, a young male cougar was shot and killed near Abitibi Lake, more than 500 kilometres north of Montreal, near the Ontario border. It was the first confirmation of a cougar in Quebec since the 1860s.

In Ontario, despite back-country suspicions that cougars still lingered in the thick forests in the centre of the province, they were widely believed to have been extirpated by the 1880s. The last official presence of the animal in Ontario was the one killed by a certain T.W. White, of Creemore, on January 4, 1884. But there is no use telling these things to David Wood of Monkland, near Cornwall. Wood swears that he was attacked by a cougar in his backyard on August 4, 2001, and he insists that he saw the cougar again the next day, in the same place. The Ontario Ministry of Natural Resources and the national Committee on the Status of Endangered Wildlife in Canada decided to remain neutral on the question, but the Ontario Puma Foundation (which prefers the taxonomically correct term for the animal) has no doubts at all. The foundation was established in the spring of 2002 to investigate the growing number of cougar sightings in Ontario and to try to scientifically prove that cougars are present in the province. Within a year, the foundation was confidently estimating that 108 cougars were living between Sault Ste. Marie and Ottawa, with an Ontario total of roughly 300 animals.

By then, cougar sightings had become commonplace throughout the eastern half of North America, from Ontario to Kentucky and as far east as Canada's Maritime provinces. Mark Pulsifer, a conservation officer whose jurisdiction takes in Nova Scotia's Guysborough and Antigonish counties, receives at least a half-dozen formal cougar-sighting reports every year. "It's always been on the verge of legend and folklore," he says. "The really perplexing part is that what people are describing is, in fact, a cougar. There is just no other mammal like that, with a metre-long tail. What I find really intriguing is that these reports often come from trained observers, RCMP officers, people like that. As a biologist, I'm really keen on putting this one to bed. It's one of those great mysteries.... It's the ghost of the woods, the elusive cougar."

At the University of Northern Arizona, Paul Beier, whose research had produced those disturbing findings about the dramatic rise in cougar attacks during the 1990s, doubted that the reports indicated a revival from relic populations of the once-abundant eastern cougar, which was long considered a distinct subspecies. Instead, Beier reckoned that the eastern sightings likely involved escaped or released pets, or their descendants, or individual cougars that had strayed from populations in western Canada or Texas, or maybe even from the nearly extinct population of Florida "panthers," which is what Floridians call their cougars. By 2004, Beier was speculating that it would only be a matter of time before a healthy, naturally reproducing population of cougars was back in the eastern portions of the continent. But people should be very careful about jumping to conclusions about an animal like the cougar, Beier warned. With a creature like this, things are never quite what they seem.

The dramatic rise in cougar attacks on Vancouver Island wasn't because cougar numbers were rising during the latter half of the twentieth century. The island's cougar population was actually shrinking. During the 1990s, it was believed to have fallen by half, to about 350 animals, owing mainly to a collapse in the numbers of coastal black-tailed deer. The drop in deer numbers appeared at least partly related to a rise in wolf populations after the 1970s, but it was also a result of changes in the island's forest cover. Logging clear-cuts had provided a bonanza in browsing pasture for deer, beginning in the 1930s, but the clear-cuts were filling in. Young stands of dense forest aren't good deer habitat, but they give cougars a distinct hunting advantage. The dense thickets create ideal ambush conditions, and so does the "edge" habitat that roads provide. But with their prey diminishing, young cougars were venturing farther afield, looking for new territory. They were getting themselves into trouble—killing sheep, chasing dogs down logging roads, and, every so often, attacking people.

Elsewhere in North America, something completely different was going on, but it was having the same effect. The dramatic rise in encounters between people and cougars in the western U.S. states could be tracked in direct relationship to human population growth and the expansion of suburbs into what was formerly prime cougar habitat. That was especially true in California, Colorado, and Washington State. Some western cougar populations likely recovered during the second half of the twentieth century, but the main thing was that more people were spending more time in wilderness areas than ever before. In the United States, it was the people that were getting into trouble.

Apart from cougar attacks, the 1990s also ended up being the worst decade in North American history for attacks on humans by sharks, alligators, and bears. A small part of the reason was that many predators just weren't as rare as they had once been, and perhaps especially in Canada, their recoveries were greeted happily. Another thing was that more people were spending more time in the predators' habitat. But the main reason was that Cotton Mather's influence on the American psyche was starting to wane. People weren't afraid of the wilderness anymore.

Even in western North America, where cougars were still a recurring threat to livestock and sometimes to children, bounties had been eliminated, by popular sentiment, by the early 1960s. What's important is that this had happened in spite of the cougar's notorious habits.

Unlike a grizzly bear, a wolf, or any other North American carnivore, a cougar will happily stalk, kill, and eat a human being. Once it chooses its victim, it follows, watches, and waits. Cougars don't hunt in packs. They rarely chase their prey. They creep, hide, and pounce. For large prey species, their method is wickedly simple, whether it's a deer, an antelope, a young elk, a sheep, a dog, a child, or a grown man, like David Parker. They attack from behind. With their huge, curved, retractable claws, they pull the victim's head to the side and slash at the throat. They grab hold with their teeth—they have long canines and immensely powerful temporalis and masseter muscles in their jaws and face—and they sever the victim's spinal column or crush its skull. It's usually all over within seconds.

Throughout the United States, the rise in cougar attacks and sightings provoked less public fear than hope for the creature's return. In California, which is second only to British Columbia among North American jurisdictions with a rising death rate from cougar attacks, voters affirmed a 1990 ballot initiative designating cougars a "specially protected mammal." It's the only animal enjoying such protection in that state. Overturning the designation requires a four-fifths vote of the legislature.

Even on Vancouver Island, the epicentre of attacks in North America, the cougars that ended up in residential areas rarely incurred anyone's wrath. Most people simply wanted them anesthetized and relocated. That's rarely the wisest or the kindest thing to do, but the point is that even with a dramatic rise in cougar attacks on the island, the people held no deep animosity toward the animal. In 1992, when a cougar found its way into the underground parking garage of the famously stately Empress Hotel on Victoria's waterfront, the incident gave rise to something approaching civic pride.

In the end, North America's cougars are survivors. They survived the wave of extinctions that carried away three dozen large North American animal species around the time that human beings first arrived on the continent. Then they survived the ring hunts, the bounty slaughters, and the vanishing of the old forests that accompanied European colonization.

Since the time of Christopher Columbus, North America has lost at least two dozen mammal species, a dozen bird species, and 30 freshwater fish species. Gone are the Texas red wolf, the Badlands bighorn sheep, the Dawson's caribou, the Penasco chipmunk, and the eastern elk. Gone are the Santa Barbara song sparrow and the San Clemente Bewick's wren. Gone are New Hampshire's silver trout, the Great Lakes blue pike, and Nevada's Alvord cutthroat trout. Gone are the passenger pigeon, the eskimo curlew, and the Carolina parakeet. Another 300 animals persist only tenuously, as endangered species.

By the end of the 1960s, North Americans were haunted by an abiding sense of loss, by bitter memories, and by fear. They clung to comforting mythologies about a lost sylvan world, trampled by the machines of modernity. Deep currents ran through the Western imagination, and one morning, in the third week of September 1971, in the waters of the North Pacific not far from Port Alice, the halibut vessel *Phyllis Cormack* passed by. It was heading north, in rough water.

The vessel had been chartered by a group of Vancouver protesters, mostly university students, pacifists, Vietnam War draft-evaders, and journalists. They were bound for the Aleutian Island of Kamchatka, to stage a seaborne demonstration against the U.S. military's plans to detonate a nuclear test bomb there. The voyage ended up being an attempt to divine some meaning in all the carnage wrought since the time of Cotton Mather and Black Jack Schwartz. The protesters aboard the *Phyllis Cormack* decided to rename their vessel *Greenpeace.*

And after all the tumults that were to follow, there were other kinds of survivors, on another island, on the far side of the world. These survivors were people, the remnants of an old Norse culture of whalers. They were the last of their kind on earth.

A whale

Drifting into the Maelstrom

Then is it much wonder that oceans so great must always be hurried to get through the gate? They have but the short time of hours. Through narrowest straits must they howl and must pitch, themselves they must torture as pagan or witch, and grumble and rumble as giants.

—Petter Dass (1647–1707), "The Trumpet of Nordland"

It was the middle of the night, and the sun was shining. At the outer edges of a violent ocean vortex that had preoccupied clerics and confounded scholars from medieval times, the ocean was as flat as a millpond. All around the cove of Sorvagen, a village on the Norwegian Sea island of Moskenesøy, nothing moved.

The night before, while the ferry *Narvik* was making its 95-kilometre passage across the broad Vestfjorden gulf, Moskenesøy and all the other islands of the Lofoten archipelago had appeared to be floating across the surface of the sea. The Swedish and German tourists on board found it all very strange. The tired Lofoteners, on their way home from shopping in the harbour town of Bodø, took no notice. It was a trick of the light. In the eerie light of the arctic midsummer, it's just something that happens. But now everything was still. The stout little fishing boats in Sorvagen's quiet cove lay placidly at anchor. There was no sound, no wind.

The craggy Lofotens, mythical abode of trolls and valkyries, rise dark and forbidding from the blue ocean, and to approaching vessels, the islands' bleak and jagged mountains routinely appear to be sinking away into the sea or withdrawing beyond the horizon. No matter what might explain that illusion, we know that the Lofotens owe their strangely gentle summers to the calming effect of the North Equatorial Current, an ocean stream that rises from the middle of the Atlantic, follows a broad, clockwise arc, and becomes the Gulf Stream and then the North Atlantic Drift. It eventually weaves its way through the dark skerries and rocky headlands of the Outer Hebrides, the Faeroes, and the Shetlands. When it reaches the Norwegian Sea, it sends outriding currents to circle within the cod-rich Vestfjorden, which separates the Lofotens from the deep fjords of Nordland's mainland coast. Even in the middle of the long, dark arctic winter, the Lofoten weather rarely falls below freezing. In the warmest weeks of summer the seas can be positively drowsy.

But it is also in these very waters that one encounters the Maelstrom, a phenomenon revealed to the outside world through the writings of the sixteenth-century Swedish bishop, historian, and geographer Olaus Magnus. In his *Compendious History of the Goths, Swedes, Vandals, and Other Northern Nations,* Magnus describes many wondrous things in the Norwegian Sea, among them a huge sea serpent, and also the kraken, a giant shipwrecking squid. On his 1539 map of Scandinavia, the Carta Gotica, the Maelstrom appears as a gigantic whirlpool just a few sea miles from Sorvagen, off the southern tip of Moskenesøy, which is the southernmost of the main Lofotens.

But the *Narvik* had no raging seas to contend with when it finally put in at Moskenesøy. It was one o'clock in the morning, and there was just a plain-looking harbour in the amber twilight. A sheer granite cliff face towered over the ferry dock, an old fish plant, and a cluster of white houses and red houses. A few fishing boats at anchor. And a cab. Its driver nodded, I climbed in the back, and we headed off to Sorvagen.

I'd booked a room there in a nineteenth-century *sjøhus,* which is a kind of fish plant, net loft, and boat shed combined. It had been converted to a bed and breakfast. That's the kind of accommodation available to visitors on the Lofotens. It's either a converted sjøhus or an old *rorbuer,* a small fisherman's shack. On the way to Sorvagen the driver asked me, in halting English, where I was from. When I told him I was a Canadian, the subject immediately turned to those other tempests that had been whirling around the Lofotens for so many years.

"Are you a Paul Watson fan?" the driver asked. Watson was the renegade Greenpeace activist who'd made a name for himself over the years by sinking eight whaling ships, two of them Lofoten whaleboats that his crews had scuttled at their moorings in the 1990s. Watson had also rammed several high-seas drift-net ships, and his Sea Shepherd Society comrades had destroyed a whale-processing plant in Iceland.

I could have told the driver that Watson and I had once been friends, and that I still held him in some esteem despite public disagreements we'd had about whaling. I could have said that I'd once spent a vacation with him on the Hawaiian island of Molokai, where we'd hunted ecosystem-wrecking feral pigs with Winchester rifles. I could have admitted that we'd actually been relieved that we hadn't come across any feral pigs and that we'd contented ourselves with shooting coconuts out of trees. I could have rambled on about how strange it is, the way such contradictions play themselves out in the way we think about other species and in the way we behave. But I didn't.

Instead, I answered the same way I had on the half-dozen other occasions I'd been asked that same question about Watson since arriving in Norway only two days before. I mumbled something about being open-minded about whaling. He seemed satisfied, and introduced himself as Oddbyrg. He was a cod-boat skipper, and he drove cab only in the off season. "Paul Watson, he is crazy," he said. I had no cause to argue with Oddbyrg. Only about 1400 people live on Moskenesøy. There aren't many cab drivers.

In the waters around the Lofotens, there are, however, a lot of minke whales.

It could be argued that the disgraceful legacy of industrial whaling did not result in the complete extinction of any whale species, but such an argument would have to rely on the narrowest meaning of the word *extinct* and on a lot of hairsplitting about the definition of the term *species*. It would certainly not tell a very true story. But at the same time, it would not be true to suggest that the world's minke whales are anywhere near extinction. Minke whales are the most abundant whales on earth. The International Whaling Commission (IWC) and the International Union for the Conservation of Nature put the world's minke whale numbers at close to a million, from several stocks of at least three subspecies.

On the Lofoten Islands, what extinction most directly threatens is a *human* society. The Lofoteners are whalers. They're minke whale hunters. Because Norway has exercised its rights under IWC rules to register an "exception" to the worldwide ban on commercial whaling, the Lofoteners are persisting in the customs of their Viking forefathers. But they have also been drawn into a swirling vortex of bitter international debates, allegations and counter-allegations, disputes and controversies.

Lofoteners are commonly accused of savagery, cruelty, and contempt for international law. Lofoteners in turn accuse their detractors of engaging in cultural imperialism and hiding an eccentric cetacean fetishism behind a mask of concern for the environment. Civil debate is swallowed in the whirlpools of the IWC 's annual proceedings and disappears within the subterranean channels of the General Agreement on Tariffs and Trade (GATT). At stake is the very effort the world is making to keep the planet's rare and vanishing species from extinction, and the arguments are threatening to cripple the Convention on the International Trade in Endangered Species. Also at stake are the commitments the world's nation-states have made to stem the cultural extinctions that are bleeding away humanity's vast diversity.

Being so close to that Maelstrom, it is no small paradox that on the island of Moskenesøy, people are living in demonstrably sustainable ways within the wild world around them, and the old ways of life carry on.

On the North American side of the North Atlantic, where the Grand Banks cod stocks were once so thick they were said to have slowed the passage of sailing ships, 99 percent of the cod are gone. But around the Lofotens, where the world's great cod fisheries had begun centuries before, the seas were still seething with cod. Lofoteners also fish turbot, halibut, plaice, and herring. They catch capelin, shrimp, lumpfish, and monkfish. But it's the cod, still, that turn spring into summer, and summer into fall. On Moskenesøy, Flakstadøy, Vestvagøy, Gimsøy, Austvagøy, and the few other inhabited islands in the Lofoten chain, fishing villages cling to rocky little harbours. There are cod-drying racks everywhere. They are ancient, odd-looking structures, built with long poles fastened to stilts with nails and twine, and they are still being assembled and taken apart every year, in time with the cod seasons, in more or less the same way the islanders' Viking ancestors did. Every year, Lofoteners hang roughly 16 million kilograms of cod on 400,000 square metres' worth of drying racks.

Fishing boats bob at their moorings at every nook in the rocks. The people live in boxy and brightly painted little houses, all jumbled together in coves of pleasantly cluttered wharfs and net sheds on pilings. Impossibly dramatic granite peaks rise straight up from the depths of the ocean and tower over everything. At the height of summer, the islands are lush and green and alive with wildflowers and teeming with kittiwakes, puffins, black guillemots, and cormorants. Waterfalls tumble out of the clouds and down through the mountain heaths, and trout run up the little creeks the waterfalls make. And in the coves, the fishing boats are pine-hulled and plump.

Reine is Moskenesøy's main village. Houses, fish plants, and docks, with a school, an old folks' home, and a gas station, all crowd around the breathtaking harbour of Reinebringen. About 500 people live in Reine. Even in the stormiest weather, a Reine fisherman can find his way home by keeping the conical massif of Olstinden off his bow. Olstinden lies behind the harbour, at the meeting of two long and narrow fjords. There is the Kirkfjord, with the great Hermansdantinden Mountain at the head of it, and the Bunesfjord, which separates Reine from the small village of Hamøy, across a narrow channel.

When I first visited Reine, among the fishing boats in the harbour were the *Reinebuen,* the *Ann-Brita,* the *Malnesfjord,* and the *Leif-Junior.* They were fitted out with small harpoon cannons on the bow. There are only two dozen such harpoon-fitted fishing boats in the Lofotens and in the Vesteralen Islands to the north, and maybe a dozen more in the mainland outports to the south. You could wonder what all the fuss was about. In the village of Reine, you would not wonder for long.

I'd decided to walk from Sorvagen to Reine rather than call Oddbyrg for a ride. It was a good seven-kilometre stroll along a winding seacoast road with a short hitchhike stretch of tunnel. But I saw no sign of the *Trøndergut,* the boat that belonged to Bjorn Hugo Bendiksen, one of the whalers I'd arranged to meet up with. It turned out that he was still out fishing, filling his quota of Greenland halibut. So I decided to stop in and see Rune Frovik, the secretary to the whalers' organization, the High North Alliance.

I found the alliance headquarters just off Reine's main street, beside the school, in a former post office that it shared with a foot-massage parlour, which does a brisk business in the summertime, when the Lofoten peaks are crawling with climbers and hikers. Frovik was sitting at a desk littered with newspaper clippings, unanswered telephone messages, and file folders. He had just come back from the annual meeting of the IWC, in Berlin, and he was in a bit of a funk.

The agenda had been dominated by an Australian initiative to push the IWC even further away from its original mandate, which was to regulate whaling in a sustainable way. The meeting was just another milestone in the IWC's transition from a rubber-stamp agency for any and all whaling to an agency controlled by forces that weren't prepared to allow whaling of any sort. The minority of pro-whaling member states, including Norway, said the Australian plan would reduce the IWC to a mere research institution with some special regulatory interests in whale-watching. Norway and the IWC's other pro-whaling members lost the vote, 25 to 20.

"It was ridiculous," Frovik said, flipping through the papers on his desk.

Frovik had grown up in a small farming and fishing village south of Bergen and had ended up in whaling advocacy only after a stint as an exchange student in Paris. It was 1993, and Norway had just resumed commercial whaling, over the IWC's objections, after a four-year hiatus. The decision had stirred the animal rights' activists whose offices were situated, inconveniently, right beside Frovik's apartment, in the fourth arrondissement.

Frovik said he couldn't bear the way misinformation had become the common currency of the world whaling debate. "I would say, please, have an opinion, but have an informed opinion," he recalled, "but it did not matter." He readily admitted that he'd found himself deeply offended by the way Norway's small-boat whalers were being demonized and caricatured as cruel and unthinking savages. "There was an injustice being done," Frovik said, "and I couldn't stand for it. In war, the victors write the history, they say, but the history the victors are writing now, I cannot stand it."

Frovik decided to write his thesis for a political science degree from the University of Bergen on whaling. At first he amused himself with endless jokes at his own expense about what a dumb career move it was, but he ended up as the High North Alliance's chief researcher, propagandist, and animateur. In the meantime, the alliance had gone from a 1991 effort between the Lofoten Regional Council and the Nordland Small-Type Whalers' Association to a broad coalition of marine mammal hunters from communities throughout the North Atlantic. Among them were Faeroese pilot whalers, Inuvialuit whale and seal hunters from Canada, Icelandic minke whale hunters, and Greenland Inuit, whose traditions of hunting whales and seals went back thousands of years.

When Frovik moved to Reine in 1996, he was 27 years old and newly married to Fabianna, a woman he'd met in France. Six years later he was the father of three young children. Blessed with an acute sense of the absurd and a quick, withering humour, Frovik immersed himself in his work. Among his duties was the business of editing and writing for the whalers' newsletter, the *International Harpoon,* which is what you'd get if you combined a scholarly periodical about cetacean research with a whaling-industry journal and Britain's *Private Eye* magazine. If Paul Watson was the world's most vigorous anti-whaling activist, Rune Frovik was Watson's polar opposite.

Watson ran away to sea as a boy, and was sailing with Greenpeace from its earliest days in Vancouver. But he left Greenpeace, striking out on his own in 1977. He saw his constituents as the whales, and he made no apologies for making his life's work the business of defending whales and seals against those who would kill them. It was as simple as that. He'd reiterated the point before I'd left for Norway. We'd had a pleasant visit aboard his converted North Sea oil-rig tender, the *Farley Mowat,* in Victoria harbour on Canada's west coast.

The *Farley Mowat,* named in honour of the Canadian writer and conservationist, was an especially ferocious-looking ship that Watson had been putting to use in perfectly creditable high-seas ambuscades of sea-turtle poachers and longliners fishing illegally in the waters around the Galapagos Islands. I'd excoriated Watson in the pages of *The Globe and Mail,* Canada's national daily, for bringing down the wrath of the world upon the Makah tribe of Washington State after they'd declared their intention to resume their whale-hunting traditions. Still, he was perfectly civil when we met.

The Makahs' right to hunt whales was enshrined in an 1855 treaty, and they'd decided to exercise that right only after nearly 70 years of abstinence. The west coast's migratory grey whale population had recovered from the brink of extinction to its pre-industrial abundance of at least 20,000 animals. The IWC agreed to let the Makah take 20 grey whales over five years, which, whether anybody liked it or not, made the Makah plan eminently sustainable, maybe even honourable. The Makah killed one whale in 1999 in an emotionally charged and widely publicized event, and then found themselves stuck inside in the notorious funhouse mirrors of the American court system. Watson was delighted with the ways things had turned out.

To give him his due, Watson had treated the Makah the same way he'd treated other whaling nations, and he was no sissy. He'd been chasing foreign trawlers off the Grand Banks in the Atlantic long before the Canadian government found virtue in the practice. I had reckoned he was wrong to vilify the Makah, and wrong to go after Norway's small whaling fleet, which was causing no harm to whale stocks. Watson thought he was right, because to him what was wrong was the killing of whales. Period. End of story.

Frovik was similarly unapologetic. His constituents were the last whaling cultures of the North Atlantic, and so long as the harvests were sustainable, he reasoned, the whalers shouldn't be forced to defend their customs and traditions against the irrational and unjustifiable demands of ill-informed outsiders. The whaling and seal-hunting peoples associated with the High North Alliance were no less entitled than anyone else to engage in sustainable harvests of the wild resources around them.

The whalers might not be as glamorous as the heroic Penan tribes of Borneo, whose struggles against rainforest clear-cutting had inspired environmentalists the world over. They were certainly not so exotic as the Kaiapo of Brazil, whose leader, Paiakan, had earned such demi-celebrity that he travelled with Sting and his blessing was sought for the marketing of Rainforest Crunch, one of Ben and Jerry's ice cream flavours. But whaling cultures, Frovik reasoned, were no less entitled to continue in their traditions.

Australians killed hundreds of thousands of kangaroos every year. New Zealanders killed at least as many possums. North American hunters shot bears, deer, moose, and other game animals in similar numbers. Every year, all over the world, millions of cows and pigs and chickens spent their lives in the hellish confinement and misery of feedlots and meat factories. Frovik reckoned that such a world was not entitled to hector and harry Norwegian whalers about their traditions. Period. End of story.

Some things can be said about whaling that are without controversy, and one of them is that minke whales are abundant mainly because they were never targeted by the world's industrial whaling fleets. The species is notoriously difficult to hunt. Minke whales don't travel in huge pods, but tend instead to lead solitary lives. They don't "blow" a shot of steamy breath like other whales. They have only a slight dorsal fin, so they're hard to see from any distance, especially in choppy water. They're also small as whales go, rarely exceeding eight metres. The

North Atlantic subspecies, *Balaenoptera acutorostrata,* has at least two stocks. The Northeast Atlantic stock is estimated to exceed 100,000 animals, the Central Atlantic roughly 70,000. Lofoten whalers were taking a few hundred minkes from these stocks every year.

Minkes are rorqual whales, which is to say they're the kind with those distinctive expanding pleats along their throats that allow them to gulp enormous volumes of water, which they strain against their baleen plates for krill and copepods and small fish. Minkes are also known to target large species of fish, such as cod. The Antarctic subspecies, *Balaenoptera bonaerensis,* is by far the most numerous, at roughly 750,000 animals. Small North Pacific stocks of the subspecies *Balaenoptera scammoni* persist, mainly in the Sea of Japan and the Sea of Okhotsk. There is also a "dwarf" minke whale, which may be a separate subspecies or may be just a form of the subspecies that occurs in the North Atlantic. It's routinely sighted from the Great Barrier Reef in Australia to the Atlantic coast of South America.

Scientists make no claim to understand much about the migratory behaviour or the social life of the North Atlantic's minke populations. The whales appear to prefer calving in southern waters during the winter months and have been found as far south as the coast of Senegal. The rest of the year they tend to spend in arctic waters, although they are occasionally sighted in the Mediterranean Sea and a juvenile was once found stranded on Georgia's Black Sea coast. What the Lofotoners say on the subject is that female minke whales start appearing in the Vestfjorden in the early spring, and after a few weeks the males begin arriving, heading north along the coast, followed a few weeks later by juveniles.

In the Lofotens, the minke is known as the *vågehval,* the bay whale, owing to the fact that it was in shallow bays and coves that the Vikings trapped minke whales during the early Middle Ages, killing them with poisoned-tipped arrows. This was still the preferred method along Norway's northern coast prior to the advent of motor vessels and harpoon cannons. It's the sort of thing people in this part of the world have been doing for thousands of years.

What Watson and Frovik would find absolutely nothing to argue about is that the history of industrial whaling is a grim narrative of the serial depletion and extirpation of whale stocks. Whenever new whaling grounds were discovered, they were mined of their whales and then abandoned. As soon as one species became so scarce as to be next to impossible to find, the industry moved on to the next one.

The world's first high-seas whalers were almost certainly the Basques, a mysterious people who speak a language related to no other on earth. Their homeland lies in the western Pyrenees, between France and Spain, and on the coast of the Bay of Biscay. Basques had long been skilled mariners and fishermen, and by the eleventh century, they were setting out in small boats, armed with long lances, to hunt the whale they called the *sardako.*

The sardako was the same whale the Norsemen called the *nordkaper,* and later English whalers called it the "right whale," because it was literally the right whale to hunt. It was slow-swimming, its blubber contained a lot of oil, and it floated when it died. By the early fifteenth century, Basques were following the sardakos as far as Greenland and Iceland, and by the sixteenth century they were hunting whales from shore stations on the coast of Labrador and Newfoundland. It was a profitable business. The Basques marketed whale oil and baleen for a variety of products, and the meat was tasty. Because it was considered fish, Catholics could eat whale meat on Fridays and on all the other holy days of the liturgical calendar.

By the early seventeenth century, commercial whaling had been taken up by many European nations, but the main players were the British and the Dutch. Other species were being hunted, particularly the North Atlantic's bowhead whales, then known as Greenland whales. Around the same time, American colonists got into the business, hunting right, grey, and sperm whales. At first, the market for whale products, especially oil, seemed as inexhaustible as the whales, but it soon became evident that the seas would give up only so much.

Whalers were forced into more distant and harsher waters, into the ice-choked seas of Davis Strait, Frobisher Bay, and Baffin Bay, and eventually into the North Pacific, the South Pacific, and the Antarctic Ocean. One after the other, whale stocks collapsed.

The North Atlantic right whales that the Basques first hunted, once among the most abundant whale stocks on the planet, numbered perhaps 300 by the end of the twentieth century. A population of Atlantic grey whales once migrated along America's eastern seaboard. It is gone. A huge population of bowhead whales used to summer in the waters around Spitzbergen, north of the Norwegian Sea. One of the first whale stocks to be subjected to intensive commercial slaughter, the Spitzbergen bowheads, conservatively estimated to have once numbered close to 30,000, had been reduced to about 100 by the late nineteenth century. A century later, their numbers had not recovered. There were once perhaps a half-million fin whales in the oceans of the southern hemisphere. By the late twentieth century, about 10,000 remained.

There was a population of southern right whales that calved every year at the mouth of Tasmania's Derwent River. There were so many of them they were considered a hazard to navigation, and the sound of their breathing was enough to keep residents of the local villages awake at night. By the 1990s, the Tasmanian Parks and Wildlife Service would make excited public announcements every time a lone right whale was sighted in the vicinity of the Derwent. It rarely happened more than once a year.

The oceans around Sri Lanka were once teeming with sperm whales, just as the waters north of the Bering Strait were once teeming with bowheads. There were also several distinct, local populations of humpback whales on North America's northwest coast. By the early years of the twentieth century, the Indian Ocean's sperm whale population was a small fraction of what it had been before the whalers found them. The bowheads north of the Bering had been reduced to a shadow of their former abundance, and the eastern arctic bowheads, north of Hudson Bay and Baffin Island, were also decimated. The resident populations of humpbacks of North America's northwest coast no longer existed. The "northwest grounds" of the Gulf of Alaska lasted about 30 years until, by the late nineteenth century, right whales were so rare that it wasn't worth the trouble to hunt the few survivors. And on it goes.

By the 1850s, the sperm whale was the whalers' prime target, and Americans dominated the industry. Whaling ships from Yankee ports such as Nantucket and New Bedford travelled to all the world's oceans, from the North Pacific to the Antarctic, and from the icy Davis Straits to the Arabian Sea. At mid-century more than 700 American whale ships hunted the world's oceans. Four of every five whaling ships on the seas flew the American flag, and more than 18,000 American sailors were engaged in the business.

Australians began whaling in the early nineteenth century and soon became one of the most irresponsible of the world's whaling nations. They went through the southern right whales like wolves through flocks of sheep.

During the nineteenth century, whale products faced steadily rising competition from such commodities as coal oil, cottonseed oil, rapeseed oil, and fish oil. After petroleum was discovered in the 1860s, the world's whales might have found a reprieve, but the industry developed new markets for whale oil in the production of soap and margarine, even lipstick and perfume. Demand for baleen rose, owing to growing markets for such things as corset stays, gown hoops, buggy whips, brushes, parasols, and watch-spring filaments. Baleen prices tripled in the final few years of the century. It was a time when industry practices were undergoing a thorough transformation.

Traditionally, high-seas whaling involved rowboats dispatched from sailing ships, and whales were killed by harpoons thrown by hand. By the early twentieth century, fast steam-powered ships with bow-mounted cannons had taken over. The speed and efficiency of the new method allowed whalers to turn their attentions to the rorqual whales, which include the fast-swimming humpback, sei, fin, and blue whales. Industry innovators began developing ever more efficient means of killing and processing their prey, and an entirely new kind of industry arose, pioneered by the Norwegian sealing magnate Svend Foyn, whose operations quickly came to dominate whaling in the North Atlantic.

Suddenly, Norwegians had taken over the industry. Norwegian boats were hunting whales off Spitzbergen and the Shetland Islands, then the Faeroes, Iceland, and Greenland. Their whaling stations were everywhere. They were on both coasts of the South American continent, and their South African stations dotted the coastline all around the Cape of Good Hope. The final plunder of the southern oceans' whales had begun in earnest.

Mariners and fishermen from Norway's Vestfold and Christianafjord areas, centred at Sandefjord, on the southern coast, dominated the new industry. Their killing fields were mainly in the Antarctic Ocean. By the 1930s, 200 diesel-powered, harpoon-equipped catcher boats plied Antarctic waters, supplying 41 huge floating abattoirs. In 1931, 37,000 whales were killed in the seas around Antarctica. These were mostly blue whales, the world's largest animals, probably the largest animals in the history of the earth. During the 1937 season, roughly 55,000 whales were hauled up the stern slipways of the factory ships. Most of the ships were Norwegian. Most of the whales were blue whales.

Of the planet's pre-industrial population of about 150,000 blue whales, only a few thousand remain.

By the middle of the twentieth century, the Americans had lost interest in the dying industry, and the United States provided the impetus behind the 1946 International Convention for the Regulation of Whaling, which established the IWC. The purpose of the convention on whaling, and its commission, was to "ensure the proper and effective conservation and development of whale stocks" and "thus make possible the orderly development of the whaling industry."

The IWC was self-enforcing, with member states given broad latitude to opt out of its rules. Member states were also entitled to engage in "scientific" whaling outside IWC quotas so long as the information gathered was made available to the commission. Closed seasons and closed areas were established. The Atlantic's humpback whales were protected in 1955, the southern oceans' humpbacks were put off limits in 1964, hunting blue whales was banned in 1965, and the North Pacific's humpbacks were protected in 1966. The annual catch of the world's remaining whales was distributed in the absurdly ineffective form of "blue whale units," which calculated the catch in total yields, without reference to species or stocks.

By the early 1970s, Canada's last whaling stations were shut down, owing to the lack of whales to kill. The Dutch and the British had already left the business for the same reason. Apart from some small-scale coastal whaling and a fleet of pirate whalers hunting outside the laughing-stock authority of the commission, the only nations engaged in high-seas whaling were the Russians and the Japanese, relative newcomers to the industry.

It was an outrageous business, with huge factory ships chasing down the last of the great whales of the North Pacific. The Japanese whaling industry developed largely as a result of American incentives in the period following the Second World War. The Russians, meanwhile, got into the high-seas whaling business only in the 1950s, when the Soviet government built history's two largest factory-fishing ships, both the size of aircraft carriers, and both fitted out for whaling. Between them, the Russians and the Japanese were taking 30,000 sperm whales a year.

Then, in 1971, Greenpeace arrived on the scene, and the world was soon familiar with the story of the industrial era's extermination of the whales of the seas.

But the whaling traditions championed by the High North Alliance are part of a completely different story. The story is so different that it doesn't even acknowledge the great fall-from-grace event at the heart of environmentalism's grand narrative: Christopher Columbus's discovery of a pristine continent peopled only by environmentally aware hunter-gatherer savants, in 1492. It is a story that simply cannot be told in the language of environmentalism that was becoming such an important lingua franca in the world around the time Greenpeace was born.

In that other story, it was the Vikings, not Columbus and his sailors, who were the first Europeans to make landfall in North America. To understand that other story, it helps to know that the Vikings who settled on the Newfoundland coast at L'Anse-aux-Meadows in the early years of the eleventh century were Norse people from Greenland, who were originally from Iceland, who were, being Norse, originally from Norway. But the important thing is that all these people were whalers.

The culture that had established itself on Norway's northern coasts about 5000 years ago was among the first of the planet's whale-hunting peoples. The people hunted harp seals, bearded seals, ringed seals, walrus, beluga whales, and narwhal. Vivid petroglyph portraits depict hunts of whales and walrus from large vessels powered by several oarsmen. Some of those early people were the ancestors of the Norse, as well as the reputedly "elf-like" Sikhirtia people and the Coastal Sámi, close cousins of the reindeer-herders better known to the outside world as Laplanders.

One of the first written accounts from Norway's north coast was a letter written by the Norse chieftain Ottar in 892. Ottar described a journey he made around the top of Finnmark into the Barents Sea, in the vicinity of the Kola Peninsula, and told of killing 56 walrus along the way. Ottar was a worldly fellow and is believed to have come from the Tromsø area, about 70 sea miles north of the Lofotens. He wrote of the Sámi people to the north of his realm, and of the Finns and Danes to his east and south. He attributed his wealth to the taxes he extracted from the Finns and the Sámi and to his prowess in hunting, fishing, and whaling.

Around the time that the ancestors of the Norse first started hunting bearded seals and beluga whales on Norway's northern coast, the Saqqaq culture was emerging along the west, south, and east coasts of Greenland. They, too, were marine mammal hunters. By 2700 years ago, they were living in sturdy, square turf-and-driftwood houses, pursuing a way of life archeologists refer to as the Dorset culture. About 1500 years after that, another culture, known as the Thule people, was engaged in hunting seals, walrus, and whales along the Greenland coast, but their Thule technology was more diversified than the Dorset people's had been. The Thule people were using dogsleds, probably as a consequence of an influx of other Inuit tribes from the west.

The Thule culture appears to have withdrawn from Greenland's southern coasts by the time Erik the Red's colonists arrived in the latter part of the tenth century, because the Viking colonists weren't trading or skirmishing with any Inuit people until the fourteenth century. The Norsemen were gone from Greenland by the fifteenth century, and the reasons are still disputed. But the Inuit still inhabited Greenland when Denmark asserted sovereignty in the early nineteenth century, and Inuit people were still migrating to Greenland, from present-day Canada, as late as the 1860s. They were still whaling, too.

The part the Vikings play in the story is especially important, because the North Atlantic whaling complex today involves certain branches of the Inuit family, as we have seen, but also the descendants of a particular Viking lineage for whom whaling has been a distinctive cultural tradition. The present-day whalers of Iceland, the Faeroes, the Lofotens, and the Nordland coast can all count among their ancestors certain Norse Vikings from a society that thrived on the Lofotens 1500 years ago. In 1983, archeologists working on the Lofoten island of Vestvagøy uncovered the remains of a sixth-century Viking longhouse at Borg, adjacent to the cove of Varpneset. It is the largest Viking structure ever discovered. The cavernous, sod-roofed chieftain's palace was 83 metres long, with a ceiling 9 metres above the floor. Among its outbuildings was a 32-metre-long cattle barn capable of housing 50 cattle. The Norse sagas associate the settlement with the first immigrants to Iceland.

The Norse were distinct among the Vikings, culturally and linguistically. From at least the third century, they had their own runic written language and a common religion of sorts, and their fierceness expressed itself in its own way. The Vikings who famously colonized deep within Russia and traded throughout the Mediterranean, Baltic, and Caspian seas were Swedes. The Vikings who raided and settled England's east coast were Danes. It was the Norse who settled in the Shetlands and the Orkneys, becoming the high castes of the tribal societies they found on those islands. They spread to the Faeroes and Iceland, and then Greenland, and, later, L'Anse-aux-Meadows. The Vikings who raided and eventually settled in the Hebrides, the Isle of Man, and Ireland were Norse. Dublin was originally a Norse town, and the Vikings who in the tenth century inhabited the English northwest, establishing kingdoms in Northumberland and Yorkshire, were Irish Norsemen.

It wasn't until the sixteenth century that the Norwegians founded their famous large-scale fisheries for cod. An Italian mariner got lost at sea, washed up in the Lofotens, and eventually returned home with stories about unimaginable abundances of dried fish, precisely the kind Italians like in their cooking. This event began the industrial-scale production of *torrfisk,* the dried cod you see hanging from drying racks all over the Lofotens in the spring, as well as *klippfisk,* salted cod, which took off as a major trade commodity in the nineteenth century. The cod fishery often engaged one out every ten adult male Norwegians. The result was a workforce of highly skilled mariners.

The nineteenth century began with Norway being a Scandinavian back eddy, its major city, Bergen, boasting only 18,000 people. It ended with Norway claiming the world's third-largest fleet of ships, with 50,000 citizens working as seamen and 5000 workers employed full-time building boats and ships.

For centuries, Norwegians had been under the thumb of the Danish monarchy. They'd been bullied by the Swedes, and the old Norse dialects had been supplanted by the closely related Danish tongue. In the outlying districts, the Norwegian dialects survived, but there were protracted language wars, and it wasn't until 1885 that the revived native language, Nynorsk, was given equal status with the Danish variant.

The Norwegians are different. They insist on it. It was one of the reasons they were so ready to defy world opinion on whaling. But just as all Vikings were not the same, all Norwegians, and all whalers, are not the same.

When Svend Foyn started building his whaling stations along Norway's north coast in the 1870s, he was given no welcome by local fishermen. The locals' methods of whale killing were largely unchanged from the Viking times of the early Middle Ages. They still occasionally corralled and slaughtered minke whales the same way their Faeroese cousins slaughtered pilot whales. The locals also claimed to have developed a relationship with the whales, just as the Malay villagers had with their village tigers, the *macan bumi.* While the Norse villagers occasionally took whales, at the same time the whales were like shepherds, they said: they drove the cod toward the shore. Abundant whales were considered necessary to ensure healthy cod fisheries.

The Sandefjord-based industrial whalers dismissed these claims as silly back-country superstitions, but local villagers were adamant. In 1880, the Norwegian government agreed to a January-to-May closed season for whaling, during the time of the cod fishery. But the northern fishermen were not content, and when the cod fishery failed in the early years of the twentieth century, they rioted. Night after night their fires burned. A whaling station was destroyed. Troops were sent in from the south.

The government eventually relented. Shocked by the northerners' anger, Norway banned all whaling in the country's northern waters in 1904. The ban lasted a decade, after which new rules prohibited the killing of right whales, small blue whales, and all whales with suckling calves.

The industrial whalers met similar resistance in the Shetlands, and in Iceland large-scale whaling was banned in 1914, with limited commercial whaling allowed to resume only in 1935.

These events are not just an interesting postscript in the history of whaling. They are important events in the history of conservation, an almost-hidden history that includes those Newfoundlanders who flogged the plunderers of the Funk Island auk rookeries and the Saginaw volunteers who fought alongside H.B. Roney in the battles for the passenger pigeon in the Petoskey forests. What distinguishes the protesters of the Nordland coast from the anti-whaling activists who followed is that the Nordland militants were also fighting to protect a way of life. And many of the protesters were themselves whalers.

It is true, as Greenpeace likes to assert, that it wasn't until the 1930s that Norwegians began hunting minke whales from diesel-powered vessels with bow-mounted harpoon cannons. But that does not mean that Norwegians were not hunting minke whales until the 1930s—they certainly were. But what those early Norwegian minke whalers were doing had little in common with what the British and the

Dutch had done, or what the Russians and the Japanese and the Americans had done.

It wasn't about soap and margarine, or corset stays and buggy whips. In Greenland and Iceland, on Baffin Island and on the Faeroes, and along the northern coast of Norway, the whale hunt was mainly about food. It was about culture.

The Faeroese drive fishery for pilot whales is a tradition going back to the days of the first Norse settlers on the Faeroes a thousand years ago. Kill records date from 1584, providing what may be the world's longest continuous data record for human harvests of wildlife. The tradition remains non-commercial and closely regulated by elaborate customary laws that govern rights to the whales and the sharing of the meat. But it is a bloody spectacle. The Faeroese engage in a kind of whaling that often attracts the attention of anti-whaling activists—and camera crews.

When a pod of pilot whales is spotted close to shore, fishing vessels surround them and drive them into shallow bays, where islanders wade among them and slaughter them with machete-like weapons. A skilled Faeroese whaler can kill a pilot whale almost instantly, by severing the spinal cord. While the slaughter is relatively humane, dozens of whales are killed at a time, the whales thrash and writhe, and the sea turns blood red. Around the world, the public recoils. But it can't be argued that the Faeroese pilot whale kill is unsustainable: the Faeroese take perhaps 1000 pilot whales a year from an unquestionably healthy population, estimated by the IWC at 778,000 animals.

Although culturally Norse, the Faeroe Islands, like Greenland, are a semi-independent possession of Denmark. To make things even more confusing, Denmark is a member of the European Union but the Faeroes are not, and Denmark is a member of the IWC but the IWC doesn't consider the Faeroese pilot whale drive fishery within its ambit of authority. Pilot whales aren't on the list of whales within the IWC's mandate because, strictly speaking, they're just big dolphins.

In Iceland, the word for a stranded whale, *hvalreki,* means "godsend." Whale meat was a significant part of the islanders' domestic economy by the thirteenth century. They hunted mostly whales trapped in fjords by arctic pack ice, and their methods were similar to the Faeroese drive fishery. Nobody feels more betrayed by the IWC than Icelanders, 90 percent of whom, according to public opinion polls, believe they should be entitled to continue sustainable commercial whaling. But Iceland's relationship with the IWC has been particularly complicated.

After Iceland lifted its ban on commercial whaling in 1935, Icelandic whalers commenced small-scale commercial hunts for minke and fin whales, mainly for the domestic market. Unlike Norway, which registered an objection to the IWC's 1986 moratorium and could consequently claim to be engaged in a legal commercial hunt that was not inconsistent with the commission's authority, Iceland had agreed with the moratorium, in a parliamentary decision with a margin of one vote. What this meant was that, by the IWC's strange rules, when Iceland wanted to hunt whales it had to resort to the commission's dubious allowances for "scientific" whaling, the loophole that Japanese whalers have so controversially exploited.

When it became obvious that the majority of the commission's member states wanted the commercial-whaling moratorium to become a permanent ban on all whaling, regardless of whether certain whale stocks could be sustainably harvested, Iceland pulled out of the IWC in 1992. It eventually was readmitted, but only after a decade-long fight. Icelanders resumed a small-scale whale hunt in 2003.

"There you see Refsick. And there, Stokviken. And here is Hermansdaden."

Bjorn Hugo Bendiksen, the 39-year-old skipper of the *Trøndergut,* was running his hand over an old map of Moskenesøy on the wall of a restaurant called the Gammelbua. Bendiksen was one of the whalers I was hoping to meet up with the day I arrived in Reine, and the Gammelbua is Reine's most popular nightspot—a room with a low, timbered ceiling in an eighteenth-century trading post and dry goods store that has been converted to an inn.

Bendiksen had just got back from five days on the Greenland halibut grounds, 240 kilometres to the southwest. He wasn't especially fond of Greenland halibut, which are not halibut at all but a kind of turbot. Unless it's smoked, it's like jelly, he said. Still, the price he was getting for them wasn't bad. He'd filled his quota, about 5000 fish. He wasn't complaining.

Before setting out for the halibut grounds, he'd already filled the 22-whale quota allocated for the year to the *Trøndergut,* a traditional Norwegian trawler-style boat 21 metres in length, built in 1959. The hunting had been easy, but because of the way hunting areas are assigned, he'd had to travel to the Bear Island grounds, 320 sea miles north of Norway's north cape. "They pull our numbers out of a hat," he said, "but of course, we have never seen this hat."

The minkes from the North Atlantic's northeastern stock that migrate northward along the Norwegian coast in the spring are headed for the waters around Finnmark. Whalers prefer to hunt the minkes in those northern waters, in the summer, when the seas are glassy and languid. You can't use echo-sounders or sonar in the hunt. It spooks the whales. You have to draw from hunting disciplines the modern world had long forgotten. You have to watch, and wait patiently.

Special licences are required for any crew member handling the harpoon cannon. It's a job you train for, and you have to be a crack shot. The harpoon is tipped with an explosive penthrite grenade, a bit like a huge aluminum shotgun shell. Most whales die instantaneously; most of the rest are dead within six minutes. Government inspectors are on board on every whaling trip, and they record, in minute detail, the circumstances of every kill. It's not a particularly pleasant business, Bjorn said. Killing animals never is.

When a whale is killed, it's hauled aboard and butchered, and its meat is stowed in the hold. To the great consternation of the whalers, they have to throw the blubber overboard. There's no market for the blubber in Norway. The Japanese would happily buy it, but the IWC's rules make it impossible for Norway to sell it on the world market.

Bjorn had made two trips to the Bear Island grounds that summer, two weeks each time. He sailed with his usual three-man crew, which includes his father, Arne Bendiksen. Arne was the skipper of the *Trøndergut* before he retired. But he never really did retire. Every year, he'd carry on about missing the life, then he'd say, well, I'd just get in the way, but maybe, yes, just one last time. This year it was no different. He'd kept Bjorn guessing until the last moment. The morning the *Trøndergut* was set to leave, he showed up at the dock. Arne likes to handle the helm.

Bjorn turned back to the map.

"Here is Horseid," he said. "Here is Kvalviken. It means 'the whale's bay.' Actually, there is a stranded whale there, right now."

It was from those small villages on the map, the villages on the outer coast of Moskenesøy, that many of the Bendiksens' ancestors had come. A lot of the old stories Lofoteners tell are from those places. The stories are often pretty grim.

We shared a meal of cod tongues, potatoes, and roasted minke whale. The meat was lean and delicious, a bit like venison, a bit like filet mignon, and not at all fishy. Over dinner, Bjorn told a story about one of his great-grandfathers, who had lived to be 105 years old and was known for his huge hands. He came from a small village on the Bunesfjord, the fjord that separates Reine from Hamøy. In the old man's village, the ground was so stony that the topsoil for the gardens had to be carried down from the mountains.

People lived in the outer-island villages because they were close to the good fishing rounds. Another reason was that to live instead in the more established villages, such as Reine, was to live like a serf. In Reine, the squires were the Sverdrups, the family that built the trading post that had become the inn with the Gammelbua restaurant. The Sverdrups still owned the place. But in the latter half of the twentieth century, the squires had gone through rough times like everybody else. During the 1980s, everybody's cod quotas had been severely reduced. This was during the moratorium on whaling, which everyone considered unnecessary and doubly calamitous. The reductions in the cod catch were necessary, though. They were the reason the Lofoteners, unlike their Canadian counterparts, still had cod. The sacrifice had been necessary, but the Lofotens were still recovering.

"In some of those villages, there are no men in the graveyards," Bjorn said, "because they would always die at sea. But in the summertime, they are the most beautiful places in the world."

One of those places was the village Bjorn's grandmother came from. The name of the village was Hell. It is empty now, but it was once a small, hardscrabble settlement at the southern tip of Moskenesøy, perilously close to the Maelstrom. The winds were said to have been so powerful that they rattled the very doors off their hinges. Bjorn's grandmother told a story once about hearing fishermen screaming in the waves before they drowned. She didn't talk about her childhood much.

We talked late into the night. We talked about the Maelstrom, and about the deep and fathomless currents just below the surface of the international whaling debate. It was part of the same conversation.

Olaus Magnus, the sixteenth-century bishop who considered the Maelstrom part of the same conversation with giant krakens and sea serpents, wrote that the Maelstrom was a thing of even greater fury than the whirlpool of Charybdis, in the Mediterranean Sea. The Greeks attributed that phenomenon to a daughter of Poseidon and

Gaia who had been turned into a monster by Zeus and was confined to a cave in the vicinity of the Straits of Messina for eternity. But Magnus preferred to attribute the Maelstrom's furies to purely natural causes. He proposed that it was the result of the ocean falling into an abyss, a sea-floor chasm "so deep that it may suddenly and promptly swallow the seafarer."

One notable seventeenth-century intellectual who pondered the Maelstrom was the German Jesuit polymath Athanasius Kircher, the author of more than 40 books on philosophy, natural history, and other subjects, fascinations that led to the rumour that he was a practitioner of the dark arts of necromancy. By the time Kircher turned to the Maelstrom, he had somehow convinced himself—and later his legions of readers—that it was the result of a sea-floor aperture that opened out into a series of subterranean channels leading to the Barents Sea and the Gulf of Bothnia in the Baltic Sea.

Just how the Maelstrom ended up whirling furiously in the European imagination as a word describing a state of chaos and uncontrollable fury appears to be a simple matter of cartography. The whirlpool off the southern tip of Moskenesøy that churns its horrors within screaming distance of Hell is known as the *Moskstraumen* in the Old Norse language that Lofoteners speak. The word *maelstrom* appears to have turned up in English via sixteenth-century Dutch marine charts.

In the "rational" and modern view, the Maelstrom can be attributed to the confluence of several geophysical factors. Oddly, it wasn't until the 1990s that anyone attempted a thorough scientific analysis to explain what it was. That was an effort by scientists from the University of Oslo's mathematics department, who employed something called a high-resolution depth-integrated tidal model to look at what made the Maelstrom a thing of such fury.

A strange bathymetry is at work around the Lofotens, and strong currents and sharp tides. When the tide floods, the waters of the Vestfjorden roar across shallowing waters, past a clockwise-rotating eddy just off the Lofottoden, as the southern tip of Moskenesøy is known. Another eddy rotates in a counterclockwise direction, between the Lofottoden and the small islands of Mosken, Vaerøy, and Rost. When the tide ebbs, the sea heads in the opposite direction, past the same warring eddies. The conflicting rotation of the eddies by itself isn't enough to do the sorts of things that confounded the scholars and clerics who pondered the Maelstrom's powers. But when the strong northward-moving current on the Lofotens' outer coasts come up against opposing storm-force winds and a strong tide pours out from the Vestfjorden, things can get out of hand pretty quickly.

In the Lofoten folk tradition, the two eddies in the waters off the Lofottoden are caused by a whirling within a cauldron, stirred by Vaerøyman and Moskeneswoman. In that story, Vaerøyman became bewitched by seven sisters who were moving across the waters of the Vestfjorden, searching for their moon-crown. Vaerøyman went to their aid, and from that night forward, whenever the full moon reminded Moskeneswoman of what she had taken to be her husband's infidelity, she stirred the cauldron of the sea with her ladle, in a counterclockwise direction. In response, Vaerøyman stirred the sea with his pole, clockwise, and that's where the opposing whirlpools of the Maelstrom come from.

"Most ordinary people have a rational view," Bjorn said. He was talking about whaling now. "They have been told whales are almost extinct, so that is why they oppose it. But it is all about whether the animal you are hunting has sufficient numbers. That is what it should be about."

There are "rational" arguments that can be mustered to make a case that whaling should be brought to an end worldwide. Whales are migratory. The seas belong to no one and everyone. It's world opinion that matters. There's nothing you can get from a whale that can't be got from something else. The IWC has a horrible record as a regulatory agency and can't be trusted to manage commercial whale hunting. Global markets cannot be regulated closely enough to ensure that only whales from "sustainable" hunts make it onto store shelves.

But for each of these arguments, there are strong and rational counter-arguments. "World opinion" is dominated by Euro-American attitudes about whales, which have been unreasonably influenced by half-truth, sentiment, and misrepresentation. The IWC can be transformed, and in fact it has adopted strict, cautious, and extremely conservative new methodologies to assess whale stocks and to set sustainable harvest limits. With advances in DNA "fingerprinting," whales can be carefully tracked from the seas into the smallest corners of the marketplace.

In the riptides and warring eddies of the whaling debate, a purely "rational" view isn't easily discerned among the flotsam. Reason and passion are not easily drawn apart in the debate, just as culture and nature are not so easily segregated in the environment. The idea of an almost mystical order of cetacean intelligence, for instance, presents a special problem. As environmentalism became entrenched in popular culture, the idea burrowed deep into the Euro-American imagination.

The notion that whales are different from other animals, that they are blessed or cursed with humanlike attributes, may have some ancient roots, but its environmentalist iteration began in the 1950s with American neurophysiologist John Lilly's studies of dolphin brain size, behaviour, and communication ability. Using brain-implanted electrodes, oscilloscopes, and high-speed sound recorders, Lilly gathered data from several dolphins. At first, he was prepared to propose that the data contained evidence of a gift for mimicry of the kind found commonly in parrots—those birds beloved of everyone from Amazonian children to pirates and Qatari billionaires. By the 1960s, though, Lilly was drawing ever more bold inferences from his data. He ended up suggesting the possibility of private conversations between dolphins, elaborate vocabularies and semantics, and something akin to telepathy. Close scrutiny of Lilly's work usually ends there, which is too bad, because he soon turned to experiments on himself and on other human subjects, using sensory-deprivation chambers and LSD to test his theories about out-of-body experience and contact with extraterrestrial life forms.

Studies of whale communication on the British Columbia coast, in 1970, involved a variety of experiments. One involved close observations of killer whale pods, using state-of-the-art technology to determine which the whales preferred—recorded music or live performances by the rock band Fireweed, from the deck of a 16-metre sailboat. This was during those heady days when American astronauts were beaming images of the great blue globe of earth back from space. Ideas about the possibility of extraterrestrial intelligence were moving from the fringe to the realm of respectable discussion.

In 1973, the immensely popular American astronomer, television personality, and environmentalist celebrity Carl Sagan saw in the American space program a heightened urgency for research about cetacean intelligence. Sagan argued for stepping up attempts to communicate with whales and dolphins in order to test the human capacity for a possible encounter with super-intelligent extraterrestrials. By learning to communicate with whales, Sagan wrote, we might better enable ourselves to detect "a message from the stars." This in itself provided a scientific purpose for making peace with whales, in the spirit of what he called "friendship and reverence, brotherhood and trust."

It was all very trippy.

Lofoten Islanders, too, have their idiosyncrasies. Like other Nordlanders, Lofoteners have been known to go positively weepy over eider ducks. They can't understand how anyone could harm an eider duck. They have been known to build little stone houses for the birds, where they might nest, and where people can harvest the down after the chicks fledge.

On the Lofoten island of Skrova, over another meal of whale meat, this time raw with a bit of soy sauce and wasabe, I listened to Ulf Ellingsen, the manager of Skrova's whale-processing plant, wax eloquent about killer whales. It would be reprehensible to kill a whale so magnificent, Ellingsen declared. The very idea was contemptible. Minke whales are different, he said. They're lovely creatures, but the way they blindly follow their ancient migration routes, grazing krill and swallowing the occasional cod, they're really no better than cows, he said. One might wonder what a Hindu brahmin would make of the analogy. But we all have our idiosyncrasies.

In Germany, public opinion polls show that most people think it's perfectly acceptable to hunt deer but find the idea of killing waterfowl thoroughly objectionable. Scandinavians consider the Australian slaughter of kangaroos barbaric, and U.S. law prohibits the import of kangaroo meat. Australians can't see what's upsetting about killing kangaroos, but Australia, which wreaked such carnage on the whaling grounds, ended up one of the world's most vituperative anti-whaling nations.

Human beings have always harboured these kinds of idiosyncrasies. A deep empathy for and identification with non-human forms of life can be found throughout human history, and not just in the mystical traditions of Eastern spirituality. Just as the evidence shows little support for the contention that pre-modern peoples were always as light on the ground as John Muir's "birds and squirrels," there's little evidence that the Western, Judeo-Christian tradition was ever incorrigibly possessed of an antipathy toward non-human life.

The classical authors Plutarch and Porphyry counselled great care and concern for animals, and the Old Testament contains many admonitions against harsh treatment of beasts of burden. The fourth-century Manichees, the sect that produced Saint Augustine, the fountainhead of Christianity's intellectual traditions, were vegetarians who believed that humans were not entitled to kill animals for food. In the twelfth century, Saint Francis of Assisi preached that animals had souls. It was a heresy, but his ideas had such resonance among common people that the Vatican had little choice but to make a saint out of him. The sixteenth-century kabbalist rabbis were loath to kill so much as fleas or snakes. Even the seventeenth-century American colonists, who hated wild animals so obsessively, wrote some of the first animal cruelty laws.

In 1641, the Massachusetts colonists promulgated a statute forbidding "tiranny or crueltie toward any bruite creature which are usuallie kept for man's use." The seventeeth-century Protestant ascetic Thomas Tryon was a vegetarian who wrote eloquently in favour of the idea that natural law was offended by those who would "assault and destroy" non-human life. Strict punishments were the prescribed sentences for cruel treatment of domestic animals in Nova Scotia by the eighteenth century, and in the year 1800, the English parliament came within two votes of passing strict laws against bull-baiting. For several years, the Lords and the Commons argued about the propriety of animal-cruelty laws. Cockfighting and the mistreatment of wild animals were outlawed in Britain in 1835. The poet John Keats is said to have beaten a butcher's boy for mistreating a kitten, and Charles Dickens provided eloquent and outraged evidence in the trial of a man charged with mistreating a horse.

Whatever it was that began with environmentalism in the early 1970s, it was not a love of animals. It was just another current in the maelstrom.

It took until 1986 for the IWC to put into effect the ten-year moratorium on industrial whaling that was first proposed in June 1972 at the United Nations Conference on the Human Environment in Stockholm. The temporary ban was long overdue, and it was desperately necessary. But by then whales were already the most prominent totem animal within environmentalism, and environmentalists were wielding enormous cultural influence, especially in the wealthiest countries of the industrialized world.

So, what was supposed to have been a ten-year moratorium ended up an entrenched policy. The very nations that had most ruthlessly hunted whales—the Americans, the Dutch, the British, and the Australians—ended up being the nations most adamantly opposed to any commercial whaling. It was as though some kind of guilt transference was going on, with the Norse and the Japanese required to do penance for the sins of the nations that had so ravaged the world's whales. As if to come to terms with the sins of their Puritan ancestors, Americans were leading the charge against the world's whaling peoples, among whom were the Norwegians and Icelanders who had taken the first measures to protect whales, including riots and outright bans on industrial whaling.

Among the many deep and conflicting currents pulling environmentalism into such uncharted and dangerous waters at the time was a recrudescence of the "noble savage" mythology that had so obscured John Muir's view of the majestic Yosemite. To invent the "Indian" in the place of that vast diversity of pre-Columbian civilizations, cultures, and ways of life, an "authentic" aboriginal way of life had to be cobbled together. And just like those arbitrary distinctions between the wild and the tamed and between nature and culture, the IWC's distinctions between "aboriginal" and everybody else have ended up producing absurd and unhelpful results.

The IWC set out its doctrine on aboriginal subsistence whaling in 1981 as "whaling for purposes of local aboriginal consumption carried out by or on behalf of aboriginal, indigenous, or native peoples." The Inuit Circumpolar Conference complains that the doctrine forces its people to "live out the fantasies of white people about Eskimos."

Greenland Inuit have protested that the IWC treats them like museum specimens, denying them the right to sell some of the meat from the whales their hunters kill. It costs money to meet the commission's expectations with regard to humane kills, and research and monitoring to ensure sustainable hunting in the modern world also cost money. "Where do people think this money comes from?" they asked. Inuit hunters pursue whales from small vessels, with hand-thrown harpoons and high-powered rifles. Sometimes they take beluga and narwhal from skiffs, and sometimes from kayaks. But mainly, they hunt minke and fin whales from fishing vessels like the Lofoteners use, with bow-mounted harpoon cannons. They're not allowed to sell any of what they kill, but much of the meat and blubber produced by the Greenland hunt ends up on retail store shelves. The IWC turns a blind eye.

In 1982, Canada withdrew from the IWC entirely. The Canadian cousins of the Greenland Inuit hunt narwhal and beluga, and to a lesser extent bowhead, under the authority of the Inuvialuit and Nunavut governments, according to Canadian federal laws rather than under the IWC's aboriginal-hunt rules. It's easier that way.

Aboriginal communities that want the IWC's blessing for subsistence hunts are expected to show a long-held and "continuing" dependence on whales. In the commission's application of this aspect of its doctrine, Norse whalers are excluded, even though the Norse are no less indigenous to the coastlines and islands where they live than the Greenland Inuit are to theirs. Japanese people are also excluded, apparently because traditional Japanese whaling was a custom that had been integrated into a cash economy of sorts for at least 1500 years. But even when Japan proposed a limited, non-commercial minke whale hunt and spent four years pleading its case, the IWC still turned it down.

On the other side of the world, the Makah people had not killed a whale in almost 70 years, so they had not maintained the necessary "continuing traditional dependence on whaling." But because the Makah were considered "Indians," the IWC agreed to allow them to kill a few grey whales every year. Then the Makah found themselves up against Paul Watson and an immensely hostile North American public. To its credit, Greenpeace sat the whole thing out, being too busy with far more important concerns. The Makah plan ended up grinding away in the slowly moving gears of the American court system.

The Chukchi people of Chukotka, on the Russian side of the Bering Sea, meanwhile, engaged in an IWC-sanctioned "aboriginal subsistence" hunt that took several hundred grey whales every year, from exactly the same stock that the Makah had their eyes on. The Chukchi hunt was mainly to supply feed for local mink farms and fox farms. "Subsistence" appears to have been read into it because the fur farms provided wages for aboriginal and non-aboriginal Chukotka people. But the area's aboriginal people had never traditionally been particularly interested in grey whales. The Yupik word for grey whale means "the one that makes you shit fast."

By the 1990s, the IWC had finally revised its whale-population assessment procedures to make them sufficient to the task of regulating at least small-scale commercial hunts targeting healthy species of whales, not just "aboriginal" subsistence hunts. But the commission's membership had become overwhelmingly dominated by nation-states opposed to any commercial whaling, no matter how sustainable.

IWC meetings had become elaborate ceremonies where two voting blocs, one led by the United States and the other mainly by the Norse cultures, engaged in ritualized argumentation. Most member states sent one or two representatives, often just a low-level consular official from the city nearest the venue for the annual gathering, but the powerful U.S. delegation routinely numbered in the dozens, with advisory staff, congressional representatives, accredited U.S. environmental organizations, and state department officials. Japan's delegation was often just as big.

Among the odd assortment of observers attending IWC meetings you could find Friends of the Earth, Friends of the Whalers, the Inuit Circumpolar Conference, Save the Children, the African Wildlife

Foundation, the Whaling Problem Discussion Committee, the Women's International League for Peace and Freedom, the Animal Kingdom Foundation, the International Transport Workers Federation, the American Friends Service Committee (otherwise known as the Quakers), and, of course, Greenpeace. There were music industry luminaries, fashion models, and sometimes even movie stars.

Between meetings, both blocs busily recruited new members. By the first years of this century, the IWC had grown from the 14 founding member states in 1946 to more than 50 nations, at least half of which had never had any direct interest in whaling. Delegates came from landlocked Mongolia and Switzerland, tiny Monaco, tinier San Marino, the People's Republic of China, and several African countries.

The commission was supposed to be basing its decisions about whaling solely on the scientific evidence for sustainable harvesting, but it wasn't. In 1993, Phil Hammond, the chairman of the IWC's scientific committee, resigned. "What's the point of having a scientific committee," he asked, "if its unanimous recommendations are treated with contempt?" The commission then moved beyond just undermining its own initial reason for being—to oversee the sustainable harvest of whales—and began undercutting the most important international covenant for species protection, the Convention on the International Trade in Endangered Species (CITES).

More than 150 nation-states had signed the CITES agreement, which was specifically intended to close world markets to products from endangered and threatened species. Sensibly, CITES depended on the IWC for information about whales. After the commission declared a moratorium on all commercial whaling in 1986, CITES automatically responded by placing all whales on its banned list.

But long after the IWC's stock assessments of the North Atlantic's minke whales revealed a robust and growing population among the most abundant whale species on earth, and even though the Norwegian hunt was unquestionably sustainable, the commission still expected CITES to keep minke whales on its Appendix I list—the list reserved for species "threatened with extinction." CITES' credibility was wholly undermined.

In 1999 CITES secretary general Willem Winjstekers protested that the IWC was essentially requiring the convention to be complicit in an act of camouflaging healthy but "special" species among legitimately endangered species. Around the same time, the anti-whaling bloc's hijacking of the IWC drew the attention of the International Union for the Conservation of Nature. Concerned that the IWC was undermining international efforts to conserve endangered species, the IUCN warned the commission that other international agencies might be required to "take decisions on whales that leave the IWC behind."

The majority of the commission's member states tried to maintain the fiction that minke whales were somehow endangered, but the IUCN refused to play the game. Unlike CITES, the IUCN wasn't bound to follow IWC advice—it instead listed minke whales in its low-risk, non-threatened category.

This is what it had come to. The "Save the Whales" lobby, where modern environmentalism was born, was threatening to corrupt the entire purpose of CITES, which is to prevent the rarest of the world's living things from being bought, sold, and traded into extinction. The great cause of saving the world's whales, once at the vanguard of the global conservation movement, had ended up a reactionary and corrosive influence in the very measures the world community was using to save endangered species from extinction's abyss.

During the early 1990s, the World Commission of Environment and Development, generally known as the Brundtland Commission, had hammered out a historic international consensus on environmental protection. The hard-won agreement had been built upon the principle that ecosystem protection and species conservation could be assured only through sustainable use of natural resources. The consensus grew out of the U.N. Conference on the Environment and Development, where the principle of sustainable use was key to winning international support for the groundbreaking Convention on Biological Diversity.

At the IWC, it was as though none of this had happened.

Gro Harlem Brundtland, the former Norwegian prime minister who had been the chair of the World Commission of Environment and Development, was unequivocal about the IWC's conduct: "We have to base resource management on science and knowledge, not on myths that some specifically designated animals are different and should not be hunted, regardless of the ecological justification for doing so. There is no alternative to the principle of sustainable development. This is necessary and logical. People haven't understood how important this is."

U.S. president Bill Clinton was unmoved by Brundtland's reasoning and came within the stroke of a pen of invoking trade sanctions against Norway for its decision to trigger the IWC's "exception" rule and resume whaling. British prime minister John Major said Norway's decision was incompatible with European Union membership. British Labour MP Tony Banks stated that as a fellow socialist, he was ashamed of Brundtland. Because Brundtland condoned whaling, "I recognize her as a murderer," Banks said. The rumpus contributed to Norway's ending up outside the European Union.

In a sternly worded 1996 statement supporting the work of the High North Alliance, Brundtland was explicit about what was at stake. It wasn't just that the IWC was being allowed to hijack international efforts to save animal and plant species from extinction. The commission was also undermining global efforts aimed at preventing the extinction of human populations and cultures. "I believe that not many people realize we are faced with a violation of other people's culture," Brundtland wrote. The IWC's intransigence was unfairly denying the world's last whaling cultures their rights of access to international markets and was also undermining international commitments to respect cultural diversity, Brundtland asserted. At stake was the right of the world's peoples to live sustainably upon the natural resources around them.

It wasn't just about whales or whaling cultures. It was about the cultural and economic survival of dozens of distinct peoples within the great boreal forests that ring the planet's subarctic latitudes. The fate of those cultures was being decided by powerful factions within environmentalism that regard the trapping of fur-bearing animals as cruel and unnecessary. It was about Latin America, where rainforests were disappearing to supply Americans with hamburgers. It was about Africa, where Euro-American ideas about the inviolable distinction between humanity and wilderness had been imposed upon the landscape in the form of huge game preserves that were sometimes causing more harm than good, for both the animals within them and the people around them.

The Forest Peoples' Programme, a European conservationist group, was finding that misery and dispossession often follow the nature-for-development trade-offs facilitated by the World Bank and other international agencies. In one case, the Exxon corporation contributed to the establishment of a massive wildlife reserve in Cameroon in 1999 as compensation for damage caused by its $2 billion U.S. Chad–Cameroon oil pipeline. The result was the eviction of the Bagyeli pygmies from their 2,000-square-kilometre forest homeland. The Forest Peoples' Programme's case studies have documented similar evictions of tribal peoples, usually small-scale pastoralists, hunter-gatherers, and farmers, as a result of new parks in Rwanda, Uganda, South Africa, Tanzania, and Kenya.

By the beginning of this century, behind all the rhetoric and controversy and mayhem, only a few thousand whales were being killed by hunters every year, and the large-scale industrial whaling industry no longer mined the oceans of whales. That's the bright side. But there is also a bright side to Norway's insistence that it would not be bullied into dispossessing its people of their whaling traditions. Norway was consciously holding out against the forces of sameness eclipsing the world. They were the same forces that were arrayed against the fiercely independent Faeroe Islanders in the nineteenth century, when even the old Faeroese language, with its strange admixture of Gaelic and an ancient variety of Norse, was as good as dead. The Faeroese people didn't even have a written literature to speak of. But the islanders hung on, and they survived, and the Faeroese language, like Hebrew, eventually defied the odds of dying languages. Like that small population of scarlet macaws at Curú, the Faeroese language came back. Today it flourishes.

There wasn't much of a "rational" economic argument for Norway's stubbornness about whaling. The Lofoteners would not disappear if Norway abandoned whaling altogether. Several hundred jobs would be lost forever, and the Lofoten economy would take a beating. Some islands, like little Skrova, where Ellingsen's plant was the main employer, would probably end up shuttered and empty. Life, though, would go on.

But the world would be diminished. It would be one more extinction. We'd all be that much closer to a monoculture, and Lofoteners would be pulled just one more degree back from a sustainable economy derived from the variety of ecological niches and abundances in the seas around their islands. Eventually, their old stories would die.

Already, in the brief summers, tour buses roll through Reine on the hour, disgorging Germans, Italians, Spaniards, and Russians. The tourists happily wander around buying mementos at little shops that had been fishermen's cabins in the days before fishing boats had diesel engines and comfortable below-decks sleeping berths. The tourists trundle back into their buses and continue on through the Lofotens, across the bridges and through the tunnels that connect the islands, and then they're gone.

It's said that on particularly stormy days, on especially strong tides, minke whales are sometimes pulled into the Maelstrom, where they die of exhaustion. When the seas around the Lofotens are calm and glassy, it is easy to forget what life on the islands is like for most of the year. The tourist season lasts 12 weeks. Then the mists set in. The rains come, and the winds start to howl.

In the wheelhouse of the 20-metre *Malnesfjord,* Hallvard Bendiksen, Bjorn's younger brother, pondered these things, while his wife, Renata, sat in the sun with their 16-month-old daughter, Anna Sophia, up on the bow. The boat moved slowly across the calm sea, heading south from the harbour at Reine. It was a beautiful day, the meat of seven whales was in the hold, and Hallvard had a couple of days to go before bringing his catch to Ellingsen's up at Skrova. We'd decided to spend a pleasant afternoon aboard the *Malnesfjord,* drifting around the edge of the Maelstrom.

When we passed Hell, the village where Hallvard's grandmother had heard fishermen screaming in the waves before they drowned, we had been pushing along at eight knots. When we reached the Lofottoden, we could feel the pull of the eddies. There was a gentle southerly swell and a light breeze, and then the pull of the current brought our speed up to ten knots. Hallvard turned to port. The *Malnesfjord* settled against the drag of the current, and we were becalmed at a place where fishermen see the *draugen,* a man alone in a broken boat, drifting. Sometimes he's headless, dressed in oilskins. He is what you see before you die. Hallvard's grandfather named his boat the *Draugen,* for luck. The boat was still fishing, still whaling.

In the *Malnesfjord*'s wheelhouse, among the depth sounders, Global Positioning System displays, electronic rudder indicator, VHF radios, autopilot, and all that other necessary bric-a-brac of modern navigation, there was a little cloth doll. Hallvard's uncle Roald, who owned the boat before Hallvard, got it from a Sámi woman. It brought good luck. "Or that is what they say," Hallvard told me. "But I am not going to remove it."

Among all the rational arguments Hallvard cited in his defence of Norway's minke whale hunt, he also noted that a whaler always leaves harbour by steering to starboard. You just do. You never take anything that even looks like a rucksack. You never use a clean harpoon line, and if it's new you have to step all over it.

There are certain kinds of cheese that you don't allow on the boat. You don't even talk about cattle. There was a government inspector who brought a magazine on board, a farming-industry magazine, and it had pictures of cows in it. Days passed, and no whales were seen. Hallvard noticed the magazine, with its pictures of cows, and threw it overboard. Then they got six whales.

Sometimes the luck is so bad a skipper has to take a flaming torch and run through every part of the boat with it. If a crow lands on the boat, that is very bad. Seagulls are the souls of drowned men; three flying in formation is an ill omen. Once, a Sámi woman wanted to buy some whale meat from Hallvard's father, but the rules prohibited it—the meat had to be landed at a government-inspected facility. But the woman was Sámi, and they have magic, and you have to be nice to them. Hallvard's father couldn't bear it. He ran down the dock and gave her some meat.

"It's not much of a stream today," Hallvard said. "It's quiet." The current slowly tugged at the hull, pulling us gently closer to the place where Captain Nemo's *Nautilus* disappears in the final pages of Jules Verne's 1870 novel *Twenty Thousands Leagues under the Sea.* It was also this very place that the narrator of Edgar Allan Poe's 1841 *Descent into the Maelstrom* watched, from a distant hill, as *the vast bed of the waters seamed and scarred into a thousand conflicting channels, burst suddenly into a frenzied convulsion—heaving, boiling, hissing—gyrating in gigantic and innumerable vortices, and all whirling and plunging on to the eastward …*

We were up to 11 knots. In the distance, a strong surface wind was pulling streamers and cats' paws across the waves. Then we saw some whitecaps. We could see little Vaerøy, and Rost, but we were not so far into the thick of things that we could make out that other island, on the far side of the Maelstrom. It's called Udrost. It is a place you can visit and not even know you've been there.

In the old days, fishermen who survived the Maelstrom would sometimes return to their little coves on Moskenesøy and discover straw wedged between the boat's rudder and its sternpost. They would cut open the cod they'd caught and find grain in the stomachs, and sometimes the fishermen returned with stories of having visited Udrost. They'd say lush fields of barley grow there, and strange flowers blossom among the stones. A man lives there, and his sons are three cormorants. The man will treat you with kindness, and if you follow the cormorants, they will always lead you to fish.

The Maelstrom pulled harder on the hull of the *Malnesfjord.* Hallvard turned to starboard and put Hell off our port, and we headed back to Reine.

A flower

An Apple Is a Kind of Rose

For the majority of humanity that does not have to watch nature on television but continues to live in rural societies, the distinctions between utility and beauty, between domesticated and wild, are much less clear.

—Pat Mooney

When my three children reached an age suitable for raiding apples from a long-abandoned orchard a short walk through the woods from our house, I did nothing to discourage them. On the island where we live, old orchards were being lost to memory, and I reckoned it was wrong to so blithely dishonour the labour of early island settlers as to let the fruit of their toil rot into the ground. Unharvested food has a smell about it that suggests something far worse than mere theft, and I'm not certain that there isn't some kind of moral duty involved in raiding fruit trees that have fallen into the possession of absentee landlords.

When my daughter Zoe turned 13, she brought home her first

boyfriend and wandered off into the woods with him. It was only when she returned shortly afterward with a bag of apples and an account of how she had put the boy to work in the orchard stealing them that I noticed the sun was still shining in the sky and the birds were still singing in the trees. My sons fell into apple stealing quite naturally. By the time Eamonn was ten and Conall was eight, they were already well accustomed to trundling off through the woods and scaling a rickety fence into the tangled and overgrown orchard to engage in the ancient, righteous, and honourable practice of stealing apples.

One evening, after my boys had returned from the orchard with an especially big bag bursting with fruit, Zoe baked me a pie. It was succulent and sweet and tart and strangely melancholy. I'd never tasted apple pie like it. We were all sitting at the kitchen table, and the kids were eating in an almost reverent silence. I asked the kids, why does the pie taste like this? After a moment, all at once, they said, it's the apples. They're different.

It's hard to explain it, Conall said. Eamonn agreed; they're just different, he said, and he added that he'd already resolved to stick with the apples he was expropriating from the orchard and to stay away from "disgusting" store-bought apples entirely.

I didn't know anything about the island's apples that would have made them "different." I did know about our tomatoes, though. Mayne Island used to have the largest greenhouse operation in the British empire, more than three hectares under glass, all tomatoes.

The island's Japanese community had built the greenhouses. The first non-aboriginal settlers were the husbands of local Native women from the Saanich and Cowichan tribes. They were Englishmen and Scots, Germans, Portuguese, and New Brunswickers, and they were good people, farmers, mainly, but the Japanese arrived as successful fishermen, and many were prominent Anglican churchgoers at St. Mary Magdalene. It was only after they got the greenhouses going that the island began to thrive.

In 1941, when the U.S. fleet was bombed at Pearl Harbor, islanders took heart. Canada had already been fighting the Second World War for two years, and the attack meant the United States would finally have to come on side. But then Canada decided that the west coast's Japanese settlers were some sort of a threat, and early in 1942 Mayne Island's Japanese were told they'd have to spend the rest of the war in internment camps. It was a disgraceful thing, and Mayne Islanders protested, but to no avail. When the ship came to take the Japanese away, the whole island turned out down at Miners Bay to say goodbye. Without the Japanese, the greenhouses petered out. Even the school closed down.

That's what I knew about our tomatoes. I didn't know anything about the island's apples except that the old orchards were being taken back by the forest. So I resolved to ask my friend Tina Farmilo.

Tina's an artist and an avocational geographer, and a couple of weeks before, I'd noticed her down at the farmer's market with a big map in a booth beside the old Agricultural Hall. She was asking people to put coloured dots on the map, to mark the island's old orchards. Tina had been surrounded by islanders and her map was covered in coloured dots, so I reckoned she might have some knowledge that would account for what had made that pie so wonderful and different. I called her on the telephone.

Yes, Mayne Island has lots of old orchards, she said. And it was once renowned for its King apples, although there probably wasn't much to the talk about the island being the first place to grow apples on the coast. Maybe Saltspring Island, Tina said. But not Mayne. The first apples in the Gulf Islands had come from such places as Nova Scotia's Annapolis Valley. They had arrived as twigs stuck into potatoes to survive the journey. The twigs were then grafted onto the islands' native crabapple trees, and that's how our first apple trees were born. Our early Kings appear to have originated in Tompkins County, New York.

I knew that the overgrown orchard where my boys had been swiping apples wasn't especially old. But maybe the trees there were the scions of Mayne Island's early King apple trees?

Could be, Tina said. I heard drawers being opened and paper being shuffled, and then she started reciting some possibilities. There were Gravensteins, Jonah Golds, and Russetts on that part of the island, Tina said. Transparents, too. And tiny little Spartan-like apples. And, oh, there was a tremendous mix of apples over at the Browns' farm. And down near Hardscrabble Farm was another orchard with Clairmonts, Early Blaise, and Red Rome. And native crabapples, those little olive-sized ones—somebody had brought them over from Saltspring. They were a favourite with the Cowichan women in the early days. Oh, we found some Winter Nellis pears the Japanese planted. Did you know Old Man Bennett had a business drying pears and shipping them to Vancouver in the forties?

No, I didn't, I said.

Yes, she said. The Century pear, she thought.

Later, Zoe reminded me that there were indeed mysterious and lovely little brown pears in the orchard. And grapes. The grapes were really odd. One day you'd go there and they weren't anywhere near ready to pick, but if you went a couple of days later it would be too late. But the apples were the best.

That telephone call to Tina Farmilo set in motion a course of inquiry in which I eventually discovered that on October 16, 1976, when Kathy Wafler Madison of Rochester, New York, was 16 years old, she created the world's longest apple peel, "172 feet, 4 inches," and that there really was a Johnny Appleseed, a Swedenborgian mystic by the name of John Chapman, and that Pliny the Elder, in AD 79, could name 20 apple varieties. I also learned that an apple is really a kind of rose, and that the evolutionary mother of all apples is *Malus sieversii,* which appears to have originated in the Tien-Shan mountains of Kazakhstan. It was the Romans who spread the apple throughout

Europe, and by the close of the nineteenth century, more than 7000 commercial varieties of apples were being cultivated throughout North America. A century later, almost all of those varieties had disappeared. A mere 15 apple varieties account for more than 90 percent of all North American production. In Canada, two-thirds of the apple crop is made up of only 3 varieties: McIntosh, Red Delicious, and Spartan. Two-thirds of the American crop is made up of Red Delicious, Golden Delicious, and Granny Smith.

All those old apple varieties, gone. In the United States, not even Thomas Jefferson's beloved Taliaferro apple, once the pride of his Monticello orchard, had survived. The Taliaferro was an apple that produced "the best cyder ... nearer to the silky champagne than any other," but it was gone forever. Gone too was the Ansault pear, once described as "better than any other pear," a fruit with a rich, sweet flavour, a delicate but distinct perfume.

It wasn't just the fruit of the labour of Mayne Island's pioneers that was being left to rot in the ground. And we weren't just losing apples. The legacy built up by small-scale farmers over thousands of years was being left to rot in orchards and fields and gardens all over the world. The result was a pace and scale of extinctions that actually surpassed what was known about the extinctions of the world's wild things.

The Slow Food Foundation for Biodiversity reckons that 30,000 vegetable varieties were lost during the twentieth century and that at the beginning of the twenty-first century, another vegetable variety was going extinct every six hours. Throughout the tropical world, bland commercial varieties of plantation bananas with little to recommend them apart from their predictable shape and shelf life were replacing ancient and succulent varieties as delightful as their names—Sweetheart, Praying Hands, Rajapuri, and A Thousand Fingers.

Until the 1990s, it was not uncommon for Thai villagers to rely on roughly 400 plant species for food and medicine. On the island of Borneo, tribespeople routinely got their sustenance from 800 different named plant varieties. The Hanunoo people of the Philippines could distinguish more than 1600 plants. The Suazi people of Swaziland cultivated and nurtured about 200 species of "wild" plants. In some remote Andean valleys, it was still possible, in the late twentieth century, to find as many as 100 distinct potato varieties under cultivation, and there were still !Kung people in South Africa who famously managed to get by in their desert homelands with the knowledge of 84 plants. The !Kung spent only two or three hours a day gathering food that provided greater nourishment than the average daily caloric intake in any other part of Africa or Asia.

But all that accumulated knowledge was rapidly vanishing. And languages were dying along with everything else. With just 5000 languages remaining at the beginning of the twenty-first century, and with the global empire of commerce and communication basing itself increasingly on English, most linguists agree that half the world's languages will be gone by 2050. Every two weeks a language dies. It's like libraries going up in flames. Historian and linguist Andrew Dalby reckons that within two centuries, there may be only 200 languages left on earth.

Languages die when they become no longer useful in people's working lives. Small languages become vulnerable when they are surrounded by big languages like Mandarin, English, Spanish, and Hindi. They die from the kind of "faunal collapse" that carries off large mammal populations confined within wilderness parks, in fragmented landscapes. Isolated languages can linger awhile in rituals, until those rituals disappear. They can linger awhile around the hearth, until the television comes on. But a dying language rarely lingers long. An ancient language can disappear from a place in the span of a human lifetime. The language of empire eventually becomes the local language. That's what was happening back in Sikachi Alyan, that village on the Amur River. Some of the old people still spoke only Nanai, and the people of the middle generation were mostly bilingual, but the children playing along the beach near the petroglyphs couldn't speak their own language. They spoke Russian.

Direct cause-and-effect lines are often difficult to draw between the vanishing of plants, animals, and language, but one thing does tend to lead to another. When "habitat loss" is cited as the main reason the world is losing all those species of mammals, birds, amphibians, and reptiles, it means that the vanishing of plants preceded the peril faced by all those other living things. ("Habitat" usually means trees, shrubs, bushes, legumes, and grasses. "Habitat loss" usually means all the trees have been cut down.) Losses in the abundance and diversity of plants are directly responsible for the imperilled state of 75 percent of the world's endangered mammals, 65 percent of endangered reptiles, 55 percent of endangered amphibians, and 45 percent of endangered birds.

The connection between the loss of vegetative cover and the fatal withering of a particular species of mammal or bird sometimes involves complicated ecological relationships—trophic cascades of the kind that befell the eskimo curlew, or "latent" extinctions of the kind condemning African primates to the ranks of the living dead. Sometimes, the connection between the extinctions of plants and animals runs the opposite way. That's what is happening to the calvaria tree, a "living dead" species on the island of Mauritius. The seeds of a calvaria tree must pass through the gizzard of a bird for successful germination; the bird it used to rely on was the dodo. Most often, though, it's the extinction of things that grow up out of the ground that precedes the disappearance of animals.

If you take the trees away in logging trucks, what's left behind tends to die. The extinction of one kind of plant also often leads to the obliteration of others, including domesticated plants, and their extinction in turn can lead to the extinction of human cultures and languages and ways of life. The borders between the wild and the tamed are surprisingly porous. All tamed things were once wild, and as with the Sonoran Desert transformed by Hohokam farmers and with the great herds of buffalo that Tim Flannery calls a "human artefact," wild things are often more domesticated than we prefer to imagine. Extinction certainly draws no such distinctions. One thing leads to another.

Roger Jimmy saw particularly straight and tragic lines between these things. Jimmy was the chief of the Kluskus band, a remote Athapaskan community in north-central British Columbia, in the hills beyond Quesnel. He didn't easily fit any "Indian" stereotype. He was conservative in many ways but was commonly derided as an Indian militant. He was both worldly and homespun. He could hunt moose and deer with the best of them and was most at ease with the old people of his community. He could also provide sensible critiques of the advice he'd gotten from Libyan president Muammar Qaddafi.

During the 1970s, Qaddafi occasionally invited North American tribal leaders to his contrarian desert republic in order to dispense counsel in the matter of how to go about the business of resisting imperialist oppression. Jimmy had once been on Qaddafi's invite list, and he found the Libyan leader to be a nice enough guy, though, on balance, ill-equipped to propose appropriate strategy in light of the objective conditions on the ground at Kluskus.

Without any outside expert advice, Jimmy had determined that by counting the kilometres of new logging roads punched into the country during the winter, you could fairly accurately estimate the number of Indian kids who would end up committing suicide during the summer. The precise cause-and-effect relationship between the numbers wasn't so easy to demonstrate, but they were definitely associated.

Similar associations have been meticulously tracked throughout the region that poet Charles Lillard called the Sitka Biome, at the epicentre of which are the Gulf Islands, the little archipelago where I live. The Sitka Biome is the ecological region more generally known as the west coast temperate rainforest. It's that huge swath of coastal redwood, cedar, fir, and hemlock forests between northern California and the Alexander archipelago of the Alaskan panhandle.

During the 1990s, the Canada–U.S. conservation group Ecotrust looked closely at the relationship between the loss of trees and the loss of aboriginal languages along North America's west coast. Ecotrust began with a strictly botanical analysis, then put that map together with another showing the status of aboriginal languages throughout the same ecological region. It ended up being the same map, only with different colours.

In the old-growth forest version, a thick blood-red line runs up from a point just north of San Francisco until it begins to thin into droplets, roughly at Johnstone Straits, at the top of the Strait of Georgia, where everything narrows between Vancouver Island and the B.C. mainland. Red means the forests are gone. North of the solid red line it's mainly green, which means there are still forests there.

On the language version of the map, 57 language areas are shown. All 19 extinct languages, with such musical names as Tillamook, Clatskanie, and Siuslaw, were spoken by peoples south of Johnstone Straits, down where the blood-red line is thickest. North of there, where the red droplets start, and farther north again, where the old forests have been ravaged in places but largely remain, the languages persist, even if only in the voices of old people. Of the 38 "living" languages in all of the Sitka Biome, 29 are spoken fluently by fewer than 100 people. The Gulf Islands are roughly in the middle of that long map, near the top end of the place where the red line is widest, where there is still some green. Among the Saanich and Cowichan people, hardly anybody speaks the old languages anymore.

You can't rely on the prism of environmentalism or on any other perspective that draws sharp lines between "wild" plants and "domesticated" plants or between a hunter-gatherer economy and a farming economy for even a glimpse of the way language loss and cultural extinction followed from the destruction of those forests. Most anthropologists, for instance, have insisted upon calling the Saanich and Cowichan people "hunter-gatherers," but the term is wholly inadequate to describe their way of life before the arrival of Europeans. It's hard to imagine why the term "farmers" should not be used to describe them, given their intensive cultivation of certain food crops.

Arrowhead is a plant with leaves like the calla lily and a tuber like a small potato, and botanists call it *Sagittaria latifolia.* It was known to the Cowichans on the Galiano Island side of Active Pass as *ska'us.* It's the same name used for the plant on the Mayne Island side of the pass among the Saanich people. Shared words were not so common between these peoples—the languages are about as similar as French and Portuguese. Across the Strait of Georgia from Mayne Island, and up the Fraser River a few kilometres, Hudson's Bay Company officials recorded that 5000 people gathered in the autumn months of 1827 to help the Katzie tribe bring in the ska'us harvest from the sloughs and marshes the tribe maintained along the Fraser. Some Katzie ska'us ponds were owned by the whole tribe, others by individual families. Ponds and sloughs were cleared in large tracts, often greater than the length of a football field and always carefully, meticulously tended.

The Saanich tribes were famous for their camas prairies. The herbaceous perennial *Camassia quamash* produces bulbs as big as potatoes, but you have to continuously cultivate them to keep them growing to that size. The Saanich harvested camas in the springtime and then dried and pounded it into flour, which they used in much the same way Europeans used wheat flour. Camas flour was kneaded into loaves and mixed with berries for cakes, and the ethnographer Erna Gunter reported in 1945 that except for choice varieties of salmon, "there was no article of food that was more widely traded than camas." The first European explorers who saw the camas prairies in the area that is now Victoria, on Vancouver Island, marvelled at the way the Saanich countryside so closely resembled the parks of southern England.

Because of these traditions, it took no time at all for potato farming to take hold throughout the Strait of Georgia, and distinct varieties emerged long before any European settlers had put down roots. The Snokomish people cultivated the "no-eyes" potato in the vicinity of what is now Blaine, a Washington State border town, and on San Juan Island the Straits Salish people were growing big, round red potatoes long before the Americans took over in the 1860s. Across Haro Straits, the Lummi people cultivated "kidney-shaped" potatoes around the same time.

All these potato varieties have disappeared, and even the once-ubiquitous camas, which produces such beautiful purple-blue flowers, was a rare sight by the close of the twentieth century. You can still find it sometimes, blooming in lonely places on the south-facing capes of the Gulf Islands, but the Saanich don't do much camas raising anymore. After most of the Saanich were swept away by the disease plagues that Europeans inadvertently brought with them, the Europeans planted other crops where the camas had grown.

Until the late nineteenth century, the many tribal peoples that had such deep roots within the west coast temperate rainforest were familiar with hundreds of species of plants that were put to use for uncounted purposes, and it's not clear what arbitrary line should demarcate those plants as "wild" rather than domesticated. The Klallam used the bark of the *sqwe e'eltc* willow for sore throats and tuberculosis. The Quinault used *t'tnixlo,* the gummy sap of the shore pine, to treat cuts and scrapes. The Lummi used *su'ktcen,* boiled vanilla leaf, as a shampoo. If they got lost in the forest, Quileute children were told, they should chew *kestola'put,* the root of the deer fern. New Makah mothers laid a certain kind of seaweed on their breasts to ease the flow of breast milk. The Skykomish carefully tended patches of red elderberries—not to eat the berries, but because deer liked to eat them, and the Skykomish liked to eat deer.

The forests go, the cultures go. As the trees, legumes, and grasses go, so goes diversity and abundance in plants, in cultures, and in languages.

John Broadhead is a friend of mine who lives on the Haida Gwaii archipelago, also known as the Queen Charlotte Islands, off British Columbia's north coast. The islands lie just southeast of the Alexander archipelago, at the top of Ecotrust's map. John works closely with the Haida people, who have almost lost their language, and he spends a great deal of time looking at maps like Ecotrust's. He also makes other maps, to chart the loss of the forests on the islands where he lives. In his work, he often finds himself recalling a line from Shakespeare's Julius Caesar: "Pardon me, thou bleeding piece of earth, that I am meek and gentle with these butchers."

It's what Mark Antony says over the body of Caesar, his dead emperor.

By the beginning of the twenty-first century, it was out of the bleeding pieces of earth that the greatest extinction in 65 million years was shedding what the Old Testament prophet Hosea had called the fowls of heaven, the beasts of the field, the land, and everyone that dwelled therein. Roughly three-quarters of the planet's old forests had been cut down, mainly during the twentieth century. Most of the world's remaining old-growth forests were in Canada, Brazil, and Russia. The United States had lost 95 percent of its original forests, Nigeria 99 percent. Countries with less than 10 percent of their old-growth forests left by the end of the twentieth century include Argentina, Bangladesh, Burma, Cambodia, Cameroon, and Sweden. Countries such as Australia, Honduras, Malaysia, and Zaire had lost about 80 percent. It is true that trees grow back, and indeed trees were growing back in many of these places, but the old forests were complicated ecologies that in some cases had evolved over thousands of years, and such lost places do not replicate themselves quickly.

In the late 1990s, the International Union for the Conservation of Nature (IUCN) reported that roughly 34,000 plants, or 12.5 percent of all the known plants in the world, were threatened with extinction. Nine out of ten of those at risk were endemics, meaning they grew in a single locale. Not surprisingly, most of the endemics were island plants, found only in such places as French Polynesia, Jamaica, Mauritius, Pitcairn, Réunion, the Seychelles, and St. Helena.

The United States ranked fourth among the world's worst-off countries in the IUCN report. More than 4600 plant species, or 29 percent of all those known to exist in the United States, were in danger of becoming extinct. What that statistic really represents is difficult to say. American data tend to be pretty robust; other countries could be at least as badly off but just don't have the data to show it. But "habitat loss" is the prime cause of the peril facing so many plants, and that usually means deforestation.

The forests are falling to make way for agricultural enterprises—usually cash crops and monocultural food-plant commodities. They are disappearing because of population growth among slash-and-burn farming cultures. They are disappearing because of climate change, as in British Columbia, where a subtle and barely measurable rise in average temperatures unleashed a plague of pine-eating beetles that were devouring a tract of pine forests the size of England. By 2003, in Alaska, spruce beetles had killed 95 percent of the spruce forests on Alaska's Kenai peninsula, over an area twice the size of Yellowstone Park. Legally protected wilderness areas and parks offer no protection against such things as global warming, and outside of them, the forests are being carted away on logging trucks.

Rational arguments can be made against opening up such bleeding wounds in the earth. One contention is that with mass deforestation we're losing untold pharmaceutical potential, that some rare tree in the Amazon may well hold the cure for leukemia, and so on. One-quarter of all known pharmaceuticals comes from forest plants, after all, and

70 percent of cancer-fighting drugs were first found in the rainforest. So we're cutting our own throats.

But just as sound is the argument Derrick Jensen and George Draffan make in their unapologetically fierce account of global deforestation, *Strangely Like War: The Global Assault on Forests.* Jensen and Draffan say the argument for saving forests for such things as pharmaceuticals isn't helping things. Rather, it is this "grotesquely narcissistic and inhuman utilitarian perspective" that got us into this mess in the first place.

It's not clear whether the aboriginal people who made their homes in the forests of the Sitka Biome would have drawn any such distinctions between utility and beauty. Every utilitarian object held the potential of becoming a work of art. From the cedars of the rainforest, west-coast cultures produced some of the most elegant and elaborate monumental art known to humanity. They used the wood, bark, and withes of cedar to create canoes, paddles and bailers, house posts, mortuary poles and heraldic poles, capes, fishing line, fish traps, hats, mats, baskets, and dozens of other things. Even halibut hooks were the most exquisite art objects.

This close proximity of utility and beauty that people find in the functions of non-human life is a netherworld that E.O. Wilson, the "father of biodiversity," has spent some time navigating. In his "biophilia" hypothesis, Wilson makes the case that human beings, over hundreds of thousands of years of evolution, have developed an "innately emotional affiliation" with other living organisms. The utilitarian and the aesthetic are inextricably linked, Wilson says, because "passion is inseverably linked to reason" in our comprehension of other life forms. Emotion is "not just a perturbation of reason but a vital part of it."

It is in that same maelstrom that one finds fourth-century Manichees and sixteenth-century kabbalist rabbis, where Russians demand protection for the Amur tigers of the Sikhote Alin mountains and Lofoteners build stone roosting places for eider ducks. It is the place where Qatari billionaires breed rare macaws and ornithologists weep at the sight of a Lord God Bird. At the vortex of that maelstrom, we disavow "tiranny or crueltie toward any bruite creature," and we are not always so meek and gentle with the butchers of the earth. The aesthetic and the utilitarian are one and the same, and a headless *draugen,* dressed in oilskins, drifts through it all in a broken boat. But of all the currents leading in and out of that maelstrom, none is more powerful than humanity's love of the things that grow up out of the ground.

Steven Pinker, an experimental psychologist best known for such works as *The Blank Slate: The Modern Denial of Human Nature* and *How the Mind Works,* argues that one of the reasons for our success as a species is our ability to recognize, value, classify, and locate distinctive forms of plant life. Our capacity to appreciate the beauty of flowers is evidence of an ancient survival strategy—a trait that evolution has specifically favoured in the human species.

The most fragrant and dazzling flowering plants also attract the greatest number of pollinators, including bees, birds, fruit bats, moths, and such primates as humans. An evolutionary side effect of the beauty of flowers was the dispersal of enormous amounts of sugar and protein throughout the world. Sugar and protein are food energy, and the abundance of herbivorous food energy, of precisely the kind found in apples, is one of the key factors that allowed for the rise and diffusion of large, warm-blooded mammal species, such as humans.

Neanderthals are known to have buried their fallen comrades with wreaths of flowers, but by Elizabethan times, flowers were necessary for quite literally pedestrian purposes: it was often impossible to walk down an English city street without a "nosegay" to ward off the foul smell of open sewers. Spices and herbs were used to mask the scent of tainted meat, and the herb garden was the pharmacy. All these years later, all over the industrialized world, people still insist upon thrusting bouquets at one another at times both happy and solemn, at weddings and at funerals, and people will go to inordinate lengths for the sake of flowers.

The case of the Kauai alula is one of my favourite examples of the lengths people will go. The alula is a species of Brighamia, a peculiar and extraordinarily beautiful succulent. It's a perennial bellflower with a rosette of sturdy green leaves erupting from a thick stem that can reach two metres in height. Its flowers are delicate white trumpets. It is one of the world's rarest flowers.

At the beginning of the 1990s, there were only 150 of these flowers remaining outside of botanical garden collections. They were confined to the Na Pali sea cliffs on the Hawaiian island of Kauai, 1000 metres above the crashing surf. A decade later, only 20 were left. Once common throughout the Hawaiian islands, the alula had come to its perilous condition because of the extinction of its only known pollinator, a moth.

To keep the Na Pali alulas alive, botanists were lowering themselves down the cliff face by rope to pollinate the flowers by hand. They were doing this every year.

It is hard to make the case that these people were behaving merely out of rational self-interest, or that they were concerned only with the potential utilitarian value of the alula. Say what you like about how horrible the human species is, appropriating 40 percent of the planet's primary productivity all to itself and chopping down all those forests. Human beings also do *this,* and they risk their lives while they're going about it.

Human beings are not solely "rational," thankfully, and there comes a point when the arguments for preserving biological diversity on the chance of finding a miracle cure in some Amazonian bug begin to sound a lot like a rationalist cover for something wholly pre-modern and deeply atavistic. They begin to sound like the words from the Apocalypse of St. John: *Be watchful and strengthen the things that remain, which are ready to die. For I find not thy works full before my God.*

They begin to sound a lot like the justifications I concoct for the pleasure I take in my own children stealing apples.

The purposes of beauty and utility in plant conservation have engaged in a particularly graceful costume-ball waltz through the years at London's famous Royal Botanical Gardens at Kew, the institution that served as the model for the London Zoo, the life's work of Sir Stamford Raffles of Singapore.

Kew is a marvellous place, and Londoners are properly proud of it. More than a million people, a good many of them Londoners, visit Kew every year. UNESCO has declared it a world heritage site, ranking it with the Taj Mahal at Agra, the Kasbah in Algiers, Cambodia's Angkor Wat, and the Great Wall of China.

Kew is fast becoming an ark for the earth's vast botanical legacy. It is 121 hectares of oddities and curiosities, hidden groves and art galleries, woodlands and sublime walks. You can get lost in the place.

Of all the old trees at Kew, the Maidenhair Tree is my favourite. It's a big old gnarly thing just a short walk from the Princess of Wales Conservatory, and the tree, as with all maidenhairs, depends on people for its very existence. Without people to cultivate it, the maidenhair would have been extinct, centuries ago. The Chinese use the maidenhair for bird's nest soup. The Japanese venerate it; Shinto traditions hold that it protected temples against fire. Oriental medicine prescribes its seeds for asthma, bronchitis, toothache, and hangovers. Both its beauty and its utility have served it well, and it's hard to say whether the maidenhair is a case of people having domesticated a plant or a plant that has domesticated people.

Kew's history can be seen as a testament to that delightful and quirky human desire to be in the presence of beauty, strangeness, and the abundance and diversity of living things. The sheer beauty of a flower, or its singularity or grandeur or oddness, was often enough to recommend its acquisition by Kew's botanists, no matter the effort.

The Temperate House, completed in 1898, is Kew's largest structure, and dominating its central atrium is a 16-metre Chilean wine palm, the world's largest indoor plant. The Temperate House is also home to what may be Kew's saddest specimen, a beautiful palm tree, the cycad *Encephalartos woodii,* which is the rarest plant in Kew's entire collection. It's the last of its kind, a lone male, from South Africa. There's also a king protea, which may be Kew's happiest plant. Nowadays, anyway. The protea was obtained in 1826, but it didn't bloom for 160 years. In 1986, it produced a huge and gorgeous pink flower, like a sunflower. It's bloomed every year since.

The Palm House is an absurdly cavernous Victorian greenhouse designed by Decimus Burton, who was also the key architect for Stamford Raffles's zoo. Completed in 1848, the Palm House is 2000 square metres under glass, and it's shaped, not accidentally, like the hull of an upturned ship. It houses more than 400 large palms, including a specimen of a cycad species, *Encephalartos altensteinii,* which is the world's oldest potted plant, having been brought to Kew in 1775.

The Palm House also features the notorious durian, a coconut-like thing that competes for title of World's Most Objectionable Plant with the fly-pollinated *Rafflesia* we encountered earlier in this book. The durian is said to be delicious but so rank-smelling that the experience of tasting it is something like eating custard in a sewer. There are also parrot flowers, papayas, fragrant frangipani, Madagascar periwinkles, a species of giant bamboo that can grow 45 centimetres in a day, and the coco-de-mer, which produces the world's largest seed.

The husks of coco-de-mer seeds were once found only on beaches around the Indian Ocean and were said to come from a magical tree at the bottom of the sea. When a Dutch fleet routed Portuguese warships that had laid siege to the Javanese sultanate of Bantam in 1602, the grateful sultan presented the Dutch admiral Wolfert Hermanszen with his most prized possession: a 23-kilogram coco-de-mer nut. It was believed to have magical properties, bestowing wisdom, health, happiness, and immunity to poison. Coco-de-mer nuts were routinely traded for fortunes, and they were commonly inlaid with gold and jewels. The nut was valued for its beauty, obviously, but far more for its utility, even though its reputed powers were, let's face it, imaginary. In 1742, French mariners found huge palm trees bearing coco-de-mer nuts on the Seychelles Islands, which rather ruined the lovely story about a magic palm tree that grew only on the ocean floor. The nuts aren't worth much anymore.

The Pagoda is a ten-storey oddity built in 1762 to satisfy the European fashion for chinoiserie in garden design. Various decorative "temples" pop up around Kew's grounds in the most unlikely places. There are the azalea gardens and rose gardens, and down through the years, the utilitarian zeal championed by patron Sir Joseph Banks, Raffles's beloved contemporary, waxed and waned in contrast with the delicate sensibilities of patrons such as Marianne North, whose 832 oil paintings of flowering plants are displayed in their own gallery on the grounds, and Queen Victoria, who donated her collection of "ordinary" plants and flowers to the Royal Botanical Gardens in 1896.

It was Banks who dispatched Kew botanists to the East Indies to fetch breadfruit to be transplanted in British plantations in the Caribbean, and the famous mutiny on the *Bounty* was the result, with sailors complaining that the breadfruit specimens on board were better treated than they were. It was also Banks who saw to it that the botanical diversity of the Sitka Biome would be known to science. In 1791, Banks issued orders to Archibald Menzies, the surgeon-botanist who would accompany Captain George Vancouver to North America's west coast in 1792. Menzies was instructed to gather either the living plants or the seeds of "all the trees, shrubs, plants, grasses, ferns and mosses" he could find. He did a pretty good job, too.

By the mid-nineteenth century, the imperial spirit and the potential economic value of the world's plant species held great sway over Kew's collections and over the activities of its botanists. Utilitarian purposes also were utmost to Sir William Hooker, who became Kew's director in 1841 and presided over the opening of the Museum of Economic Botany in 1847.

The exhibits at the Museum of Economic Botany include a spectacular array of curiosities made from plants: crude pharmaceuticals, organic record-player needles, weird hats, dentures, baskets, gourd castanets, opium pipes, snowshoes, bracelets, poisoned arrows, shampoos, earrings, toothpastes, and on and on. But the museum displays only a small portion of Kew's economic-botany collections.

Julia Steele, the assistant curator who showed me around the huge climate-controlled vault where the main economic-botany collections are stored, kept returning to the point that extinctions were taking a massive toll upon humanity's accumulated knowledge. Recording the many ways people have used plants through the ages is a key function of the museum's work. "They're so easily forgotten," Steele said. "Knowledge gets passed down from generation to generation, and if it isn't recorded properly, it can just get lost."

Steele swung open the doors to reveal a vast, brightly lit room with long columns of space-saving cabinets on wheels. The smallest bits of information can be economically important, she said, especially those things so commonly dismissed as back-country superstitions, such as the belief that a certain kind of dock plant is effective in treating the pain from stinging nettles, or the notion that putting an onion in your sock will keep tuberculosis at bay. Besides keeping track of those small bits of information, the museum also has huge things to take care of, including the collection's 32,000 specimens of wood types. Among those specimens are possible fuel substitutes or construction-material substitutes for tree species that are vanishing or are already extinct.

But as we walked down aisles and randomly opened drawers, the line between such obvious utility and beauty was not so clear. The drawers held little musical instruments, a small coffee-table sort of thing from Ethiopia made of woven grass, a Japanese sake bottle fashioned from a gourd, and a calabash pipe from the old Cape Colony in South Africa. An entire section was devoted to things made from plants while they were still living and growing. One of those things was a Chinese cricket cage that was a gourd grown in a terra-cotta mould so that it took on the mould's exact shape, complete with Chinese characters intricately pressed into its sides.

"This is really exquisite, don't you think?" Steele asked, and I agreed that it was the loveliest cricket cage I'd ever seen. She read from a note attached to it: "In climbing the sides, they fall and fight each other, striking vigorously." In another drawer, a wooden snuffbox from Ladakh, donated by a Dr. A.E.T. Aitcheson in 1873, prompted Steele to remember, "Oh, yes, I think we have some snuffboxes from Afghanistan, too," and she opened another drawer. Instead of Afghani snuffboxes, there was a collection of mulberry-tree paper that a certain Sir Harry Parkes donated to Kew after visiting Japan in the nineteenth century. "Look, it's just like silk," Steele said. "It's beautiful, don't you think?" Yes, I said. Beautiful. There in the economic-botany collections at Kew, the world's greatest collection of things made from plants for their economic utility, there was, nonetheless, all that beauty.

The tension between these two purposes is palpable at Kew, and the strange dance between them had not always been so pleasant as that visit to the economic-botany vaults with Julia Steele. Where one sees beauty, another may see vulgarity, and it was only the recommendations of a report prepared for the British parliament in 1840 that rescued 60 years' worth of plant collections from a plan to convert a series of glass houses into an indoor vineyard for the grapes at King

George III's table. And while it was then that Kew became an official repository of botanical collections, its history goes back much further than that.

Kew has been associated with intensive gardening by a variety of royals, earls, barons, and other such toffs going back to medieval times. William Turner, the "father of English botany," established a garden at Kew in the 1500s, and a century later, a certain Sir Henry Capel was busy there building greenhouses for his orange trees. In 1718, the Prince and Princess of Wales—later to be King George II and Queen Caroline—moved into the mansion that came to be called Kew Palace, on the south bank of the Thames River, after the prince got them kicked out of St. James's Palace by his father, King George I.

The banished couple were avid gardeners, but their son Frederick and his wife, Princess Augusta, were both positively barmy about gardening, as the English often are. Frederick had all sorts of outrageous landscaping plans for the place, including a replica of Mount Parnassus. Thankfully, these plans were forgotten when he died of complications arising from being struck by a cricket ball.

In most accounts, it is Princess Augusta who gets the credit for establishing the public gardens at Kew. Her obsessions were most ambitious: she set about creating a garden that was to contain a specimen of every species of plant on earth. That goal is as unlikely now as it was then, but collecting the greatest possible diversity of botanical specimens remains one of Kew's primary objectives in conservation biology.

It is a purpose that has taken on a great deal of urgency. The idea is to construct a herbarium for the ark of earth's biological diversity—an enterprise just like the work underway at Britain's Natural History Museum, where scientists are hoping to preserve the genetic blueprints of all the world's endangered species in the event that, one day, science will prove capable of conjuring extinct species from the dead.

Among Kew's collections are specimens from 15,000 separate groups of plant families. Among these are 15 species that are extinct in the wild and 2000 species at immediate risk of extinction. The herbarium contains at least some specimens from 98 percent of all the world's known species' groups, or genera. Seed banking and genetic fingerprinting are central to Kew's mandate, as is the important business of working with indigenous cultures around the world to document and catalogue plant types and specific plant uses. These things can get lost so easily, as Steele pointed out. It is a race against time.

Genetic diversity in plants is the reservoir that provides defences against the various blights, spots, viruses, leaf moulds, nematodes, and wilts that routinely afflict important food crops. Throughout the twentieth century, that reservoir was becoming dangerously shallow, owing to industrial agriculture's overwhelming reliance upon an ever-narrowing range of commercial hybrid crops. The shallowing had been going on for years, but hardly anybody noticed.

One of the first people to take these things seriously was a polymathic Russian geneticist, geographer, plant-breeder, and adventurer by the name of Nikolai Ivanovich Vavilov. He was a senior Communist Party official in the heady days before Josef Stalin took the Soviet helm, and as director of the Soviets' All Union Institute of Applied Botany and New Crops, he travelled the world collecting seeds and plant specimens. He visited more than 60 countries on his collecting expeditions, and ended up adding more than 250,000 plant varieties to Soviet seed collections.

Vavilov was once arrested as a spy on his return from a trip to Iran. He contracted malaria in Syria, fended off attacks by bandits in Ethiopia, and survived a plane crash in the Sahara. He ended up jailed by Stalin's paranoid secret police and died of malnutrition, of all things, in a Soviet gulag, in 1943. It wasn't until 1968 that Vavilov's contributions to botanical science were recognized. That year, the Soviet Communist Party authorized the issue of a stamp with his picture on it. Although his name is little known outside the small community of plant geneticists, Vavilov was recognized by UNESCO in 1987 as one of the great scientists of the twentieth century.

Among Vavilov's contributions were his discoveries about the independent emergence of civilization in epicentres of productive food-plant varieties, a subject explored in such later texts as Jared Diamond's *Guns, Germs and Steel*. Vavilov was the first to identify epicentres of food-plant diversity around the world, and while his cartography has undergone refinement over the years, he got it basically right.

Each epicentre can be identified by certain preconditions, including varied topography, aggregations of microclimates and soil types, and geophysical factors that contribute to isolation. By the late twentieth century, other fields of study were contributing to Vavilov's cartography, including linguistics, archeology, and anthropology. Epicentres of diversity are often the places where plant species originated. And it is precisely in these important places—in India, Mexico, the Middle East, Africa, and Asia—that crop monoculture is rooting out diversity the fastest, and the old crop varieties are going extinct the fastest.

By 1949, the United Nations' Food and Agriculture Organization (FAO) was already expressing concern about the need to conserve the world's food-plant seeds. By the 1970s, most of Europe's old vegetable varieties were found to be threatened with extinction. But it wasn't until 1979 that the FAO really noticed that a massive genetic erosion was occurring in the world's plant-food resources. That the erosion was showing up on international food-security radar screens at all was largely the result of efforts by the indefatigable Winnipeg Irishman Pat Mooney, who is to the cause of food-plant diversity what the great E.O. Wilson is to the whole idea of biological diversity.

By the early 1980s, Mooney, working with the acclaimed Dag Hammarskjold Centre, was warning that a "plague of sameness" was descending upon the plants of the world. His research suggested that by the twentieth century's midpoint, perhaps three-quarters of the genetic diversity in all major food crops had already been lost. It was a pandemic, and it was a grave threat to the world's major food crops, leaving them defenceless against the inevitable recurrence of diseases and blights. It was a reconstruction of the conditions present in the East Clare hills in the years before the Great Hunger. And nobody was paying attention.

The world's "gene banks" for food crops are still scattered throughout a number of disparately funded and largely neglected institutions. Little exists in the way of effective and properly funded cooperation among the various agencies that are collecting, conserving, and storing seeds around the world. Institutions such as Kew are doing their best, but we still have no comprehensive plant germplasm database. After the Soviet collapse, Vavilov's institute—where scientists starved to death during the Second World War's 900-day siege of Leningrad rather than eat the potatoes and rice in the vaults—was in a shambles. This was especially distressing, since 25 percent of the plant varieties in the institute's 330,000-entry collection had become extinct in the outside world. In 1991, after years of war and famine, Ethiopia's wheat crops were re-established only through the 3000 samples of ancient Ethiopian wheat varieties Vavilov had collected 70 years before. Even so, Vavilov's institute came a hair's breadth from being evicted, its collections scattered, when the Soviet Union's "new capitalist" government wanted the buildings for a real estate development in 2003.

The true scale of the extinctions of domesticated plants is practically impossible to quantify. The problem is much the same as the challenge conservation biologists are facing as they attempt to assemble hard data about the scope and scale of animal and plant extinctions in the tropics—the information just isn't there in any form that's easily retrieved and analyzed. But research carried out by Rural Advancement Fund International (RAFI) has gone a long way to illustrate the problem of food-crop extinctions. RAFI's findings, completed in the

1980s, are as illustrative as the 2003 Singapore study by Barry Brook, Navjot Sodhi, and Peter Ng that had focused on the scale of extinctions among "wild things" in the tropical regions of the world, the epicentre of global extinctions.

RAFI's work focused on the United States, which is as data-rich in its history of plant-food varieties as Singapore is in its records of indigenous plant and animal life. American records made it possible for RAFI's researchers to describe, quite comprehensively, the losses in food diversity suffered by a single country.

During the early years of the twentieth century, the U.S. Department of Agriculture maintained meticulous records of all the plant varieties offered by commercial seed companies in the United States. RAFI compared the early Department of Agriculture records with seed-company offerings in the 1980s and with seed holdings in the U.S. National Seed Storage Laboratory and other U.S. and European facilities. It found that about 90 percent of all the fruit and vegetable varieties American farmers were growing in the early twentieth century had vanished.

It was RAFI's research that documented the disappearance of all those apples and pears—the loss of 86.2 percent of 7098 apple varieties and 87.7 percent of 2683 pear varieties known to nineteenth-century American farmers. In 1903, American farmers and seed producers had already devised, stumbled upon, or conjured by trial and error 34 distinct varieties of Brussels sprouts; none had survived the enforced monotony that accompanied the rise of commercial hybrids. Of 544 American varieties of cabbage in 1903, only 28 showed up 80 years later in the collections of the U.S. National Seed Laboratory. Of 307 American sweet-corn varieties, only 12 remained. There were 287 carrot varieties available to Americans when the twentieth century began, but 80 years later, 21. All that remained of 357 onion varieties was a paltry collection of 21. Of 463 radishes, 27. Of 486 lettuces, 36. Of 223 watermelons, 20.

Just how all this happened is fairly straightforward. Throughout the twentieth century, North America's seed companies were dropping the old varieties in favour of fewer and fewer high-yield hybrids. The great powers of science were harnessed to construct the single most profitable hybrid—the organic equivalent of the Singapore Merlion—designed for the most efficient and effective forms of commercial production and distribution. The big prize in every type of vegetable was the hybrid with the longest shelf life, the ones that didn't bruise, were all the same size, and came ready for harvest all at the same time.

Then, one after the other, the companies themselves ended up the property of a handful of petrochemical giants that manufactured the fertilizer and the pesticides required to grow the hybrids. By the 1990s, multinationals such as Monsanto, Dupont, Dow Chemical, Cargill, Novartis, and Pioneer Hi-Bred were gaining ever-greater control over the ownership, production, and distribution of hybrid seed varieties, seed patents, crops, fertilizers, pesticides, farmland, and food commodities. Almost half of the shrinking global seed reservoir had fallen under the control of ten companies.

It's perfectly rational for the companies to offer ever fewer varieties, each designed for cultivation with the company's own fertilizers. Patented genetically modified plants are making up an ever-higher share of the companies' offerings, and their scientists are developing patent-protected "terminator" seeds that produce plants that kill their own seeds or refuse to grow at all unless sprayed with one of the company's herbicides.

While thousands of local fruit and vegetable varieties were lost in this inexorable war of attrition, the possibility always existed that some of them existed somewhere, as seeds in somebody's drawer or in a badly organized university collection or a museum repository, or maybe even in a living, growing form in some overgrown apple orchard. But still, all those American varieties—the very crops that once made the

American countryside such a varied and productive landscape—had become functionally extinct.

Beyond the United States, the data on the collapse in food-plant varieties aren't as robust, but the situation is obviously desperate. All over the world, human beings have come to rely on a surprisingly small number of plants for food. Half the world's food supply comes from grains; by the 1970s, half the wheat grown in those countries where wheat originated thousands of years ago was coming from new "miracle" varieties. Roughly 75 percent of all global food production was coming from only a dozen crops, and new, hybrid varieties were supplanting ancient and venerable varieties everywhere. In Turkey, for instance, ancient, indigenous types of wheat that were once ubiquitous were found only in the most hidden-away mountain villages.

The 30 crops that provide 95 percent of humanity's nutrition originated in Latin America, Africa, and Asia, and in these places, the replacement of old varieties with industrial hybrids is especially disturbing. These are the places where Vavilov found his epicentres of plant diversity. They lie in a narrow belt where botanical diversity flourished while much of the temperate earth was locked in the grip of the ice age.

Through the centuries, as farmers moved out into the more temperate regions of the world, they took their crops with them, and while those crops were initially few, they tended to flourish in new varieties in their new homes. Successful and sustainable plant domestication has always depended upon the profusion of varieties, along with the persistence of the plant's "primitive" and wild ancestors back where genetic complexity and resilience still existed, in Vavilov's epicentres of diversity. The wild and the tamed continued their dance down through the ages, like the dance of beauty and utility. But all the ancestor crops, the "wild" crops and the primitive crops, are disappearing. With the rapid emergence of industrial agricultural methods and the globalization of agricultural commodity production and distribution, vanishing along with all those old crops are flavours, ideas, taxonomies, pharmacies, ways of life, and customs. Like the loss of language, all it takes is a single human generation to stop cultivating an old crop variety for two or three seasons, and it's gone forever. Its characteristics are forgotten. Its distinctive strengths and its various uses are forgotten. Its taste is forgotten. The stories associated with it become extinct. The techniques of cultivating and harvesting it become obsolete, and the myths and songs that grew up around it become extinct.

For the hundreds of millions of people on earth directly engaged in farming, the twentieth century changed everything. Certainly not all those changes were for the worst, but daily life was increasingly occupied with agricultural practices that had no history at all. Farming was becoming every bit as deracinated and dehistoricized as Singapore. Even domesticated livestock breeds were disappearing. Everyone had turned to raising and breeding chimeras.

In the 1990s, the FAO reckoned that the diversity of major livestock breeds around the world was disappearing at a rate of about 5 percent a year. The bird breeds were the worst off: 63 percent were facing extinction. Half of Europe's distinct livestock breeds disappeared during the twentieth century, and 43 percent of the breeds that had survived were threatened with extinction. Half of India's goat breeds were on the verge of disappearing. The exceptionally hardy Yakut cattle of the Russian Far East had been reduced to a herd of 900 animals. The Arvana Kazahk dromedary was almost extinct, and the Philippines' Banabo chickens, which once numbered in the hundreds of thousands, were down to a few thousand birds. Only 14,000 H'mong cattle remained in Vietnam.

The FAO estimates that by the beginning of the twentieth century, about 4000 breeds of animals had been domesticated, but by the end of the century, 618 of those distinct breeds were extinct and another 475 were critically threatened with extinction. These were animals that had developed resilience to local stresses, diseases, and parasites, and were adapted to local human-modified environments. By the end of the twentieth century, the world was losing, on average, two breeds every week.

It's here that the blight that descended on the Irish potato fields at the beginning of this book shows up again. The story of the vicious little oomycete that caused it, *Phytophthora infestans,* did not end in the mass graves of An Casaoireach, that place near Tuamgraney. It did not start there, either.

Like all oomycetes, its ancestry is pretty foggy, but *Phytophthora infestans* appears to have been present in the forests of Mexico's Toluca Valley for a long time, going about the necessary work of turning various species of nightshade into rot and loam after their death. The tomato is a nightshade. So is the potato. Just when *Phytophthora infestans* turned on living, cultivated potato varieties is not known, but it likely made that great evolutionary leap at some point during the sixteenth century. That was when the Cortés family seized the Toluca Valley from its former owners among the Huastec and Otomi peoples, and turned it into a series of vast potato fields.

The first demonstration of the oomycete's ability to destroy potato crops occurred near Philadelphia, Pennsylvania, in 1843. Then in August of 1845 it jumped to Ireland and, suddenly, more than three million people—more than one-third of the entire Irish population—found themselves with almost nothing to eat. The potato crop failed all over Europe, but only in Ireland did so many people depend on the potato for their daily sustenance. Famine's horsemen rode through the Irish countryside. Before it was over, close to two-thirds of the Irish starved or fled, and the country lost almost all its speakers of Irish Gaelic. After the Irish famine, it wasn't until blight-resistant potatoes were found among the ancestral varieties growing in Mexico and the southern Andes that Europe's potato crops were restored.

The apocalyptic horsemen that had rampaged through the East Clare hills in the mid-nineteenth century were deforestation, overpopulation, crop monoculture, and vast inequities in wealth and power. A century and a half later, they were loosed upon the whole world.

The potato blight was the first time in modern history that a creature had come out of the wild and into the fields of the tamed to destroy an economically significant domesticated crop. It was also the first time such a crop was restored to abundance from genetic defences found in its distant and ancestral relatives. It wasn't the last time this would happen. In the 1870s, coffee rust wiped out coffee crops in India, Ceylon, parts of Africa, and East Asia. Cotton failed in the United States in the 1890s, and in 1904 U.S. wheat crops were devastated by wheat rust, a plague that returned in 1917. Famine broke out in Bengal when rice crops succumbed to brown spot disease in the early 1940s, and that same decade, most of the American oat crop was lost to a blight. In each case, it was genetic resistance that came to the rescue, from little-known varieties or "primitive" crop types among the plants' wild cousins back in Vavilov's epicentres of diversity. Spinach, sugar cane, black pepper, peanuts, sunflowers, strawberries, tobacco, and tomatoes have each in their turn called upon their distant, noncommercial relatives for redemption.

But the epicentres of diversity where those saviour varieties tend to be found are succumbing to Mooney's "plague of sameness" all over the world, and it was to this vulnerable world, in 1976, that the oomycete from the days of the Great Hunger returned. Again, it came out of the Toluca Valley, but by this time *Phytophthora infestans* had further evolved, and it was even more aggressive. Through the 1980s and 1990s, the potato blight showed up in Switzerland, the Netherlands, Germany, England, Egypt, Poland, Japan, Brazil, Israel, Rwanda, and then Ireland again. It appeared in Korea, Taiwan, the Philippines, and Bolivia.

The great Darwinian epoch had not ended for *Phytophthora infestans;* by the 1990s it was evolving again, mutating into ever more cunning strains. Formerly, it had been known to reproduce only by clonal propagation. The new strains were sexually reproducing. Pharmaceutical giant Ciba-Geigy developed chemical defences against the original strain, but the defences proved ineffective against the oomycete's new and wildly virulent varieties. As new chemicals and fungicides were deployed, the new strains mutated and evolved new defences and offensive strategies of their own. By the close of the twentieth century, the battle was still going on. It was costing farmers in Asia, Africa, and Latin America roughly $3 billion U.S. annually.

Scientists from Cornell University and from Poland's Mlochow Research Centre went searching with a team of Russian scientists for resistant varieties among the 10,000-entry potato collection at Vavilov's institute. Other scientists were combing the mountain villages of the Peruvian Andes. Blight-resistant varieties were being found among potato varieties still cultivated by the descendants of the very people who first cultivated the potato, 7000 years before.

Nobody was expecting science to develop a knockout punch. It was not just the diversity of food crops that was vanishing—humanity's knowledge about them was also vanishing, and to find defences against the blights and rusts and other microscopic pathogens that are necessary creatures to the functioning of life, scientists are now sifting among the rubble and ghost heaps of language, local knowledge, and old ways of life. Reviewing that search, genetic-erosion crusader Pat Mooney saw something nobody had expected to see. "Our generation," Mooney said, "may be the first in the history of the world to lose more knowledge than we gain."

The extinction of local knowledge, local ideas, and local ways of doing things accelerated rapidly during the Cold War. The Soviet and American empires were forcing the world to choose between two forms of dunning homogeneity. The United States rightly saw the need to keep the Third World's farmers willingly beholden to the capitalist side in the many proxy wars it was sponsoring against Soviet-allied forces, and even against indigenous, democratic-socialist forces.

Bold initiatives pioneered by such American institutions as the Rockefeller Foundation led to the Green Revolution, which set off an upheaval of mechanization and the mass production of hybrid food crops and uniform cash crops. It was a worldwide crusade to convert traditional forms of agriculture into the business of machines, pesticides, and fertilizers. Institutions such as the International Maize and Wheat Improvement Center and the International Rice Research Institute presented Third World farmers with the decisions they were allowed to make only as consumers and as individuals. The farmers were offered seeds, chemical fertilizers, loans, and farm machines. In return, they were expected to plant only those seeds, which required those same fertilizers and those same machines.

The revolution achieved its aims: it reduced the number of undernourished people in the Third World, and it turned many poor countries into net exporters of food. But it also touched off a rapid expansion of Third World populations and ultimately transformed millions of formerly independent farmers into vassals of the global market economy. It orchestrated a massive transfer of wealth from farmers to multinational seed companies, farm-machine companies, and fertilizer companies. And it solved local problems only by exchanging them for unprecedented, global-scale problems. As the American writer Richard Manning points out, the Green Revolution facilitated a doubling of the world's population between the 1960s and the 1990s, and it created a vast underclass of people in the cities of the world, driven from the farmland of their forebears by machines.

Manning's verdict on the revolution is that it was possibly "the worst thing that has ever happened to the planet." Apart from the extinctions of life forms, crop varieties, languages, and cultures that it caused in fairly direct ways, it was also a key contributor to the greenhouse gases that are warming the global climate and setting off such disruptions as those beetle infestations that are eating forests from Alaska to New Mexico. As recently as the 1940s, industrial farmers were using a calorie of fossil energy to produce 2.3 calories of food energy. Twenty years later, the ratio was down to 1:1.

Throughout the world, farmers increasingly rely upon seeds, machines, pesticides, herbicides, and fertilizers from somewhere else. They have found their lives being shaped and formed by decisions made somewhere else. Encoded in the new crops they are growing are the formulas by which their cultures vanish, and they have little choice in the matter. Their children are turning away from their increasingly irrelevant local customs, and all the instructions on the box telling them how much herbicide to apply to the paddy fields are written in English.

The old livestock breeds are disappearing for more or less the same reasons as so many languages and domesticated plant varieties are disappearing. The conversion to global market economies has narrowed the range of marketable livestock and poultry types to those suited for mass production and distribution. It's not that farmers want to lose their old breeds—there just isn't a market incentive to hold onto them. The big buyers in industrialized countries aren't interested in them. They are interested in mass-produced chickens from gigantic chicken factories, where the birds are kept in pens their entire, miserable lives—which are thankfully brief, owing to breeding innovations that produce chickens ready for slaughter only 33 days after hatching. In these ways, the old breeds are disappearing, lingering unwanted among local crop varieties that survive only in the cemeteries, ruined monasteries, and alleyways of remote mountain villages.

After the Cold War ended, there was only one form of dunning homogeneity to labour under. Now the promise of a new revolution is before us. It is the promise of a simulacrum of the real world every bit as convincing as the simulated reality of the Singapore zoo. It comes in the form of brighter and better chimeras, nanotechnology, transgenic animals, and genetically modified crops. It offers technologies that can make a crumpled piece of paper taste like a marshmallow, just by adding minute quantities of two simple chemical compounds. It promises that there will never be another famine. The "markets first" road is opening up horizons far beyond the hills of the Darwinian epoch, with all those bothersome little fields like a quilt, and all those stories, and all that talk about some stream that will run red with blood. It's a promise far and away more exhilarating than the arrangements Third World farmers were offered during the 1960s. It seems like such a bargain.

All the new transnationals get out of the deal is ownership of the world's food crops through patents to all their seeds, along with full control over the complex processes necessary to cultivate those crops, a monopoly over the chemicals needed to raise those crops, and the right to close down any competition. They get to decide what diversity is, and what "extinction" means. They get control over the wild and the tamed and everything in between.

If you let yourself imagine that *Homo sapiens* is a plague upon the earth and that humanity has wilfully driven birds or mammals into extinction's abyss, then it is easy to imagine that the people of Cambodia, Honduras, Malaysia, and Zaire care nothing about losing almost all of their old forests. But you would also have to imagine that the rest of us are happy to be eating the same rancid-tasting factory chickens, the same flavourless watermelons, the same predictable lettuces, and the same uninteresting, uniformly sized corn, all the time.

You would have to conclude that in 2003, India's farmers wanted to be paid 40 percent less for their wheat than they'd been paid in the mid-1990s, and that they wanted to be paid 40 percent less for their soybeans and half the price they'd been getting for their cotton. When 300 of those farmers committed suicide over a six-week period in Andhra Pradesh, you would not have seen it coming.

If you drive down the I-5 freeway that runs south from Vancouver to San Francisco, you traverse a corridor through vast and empty stretches of uniform hybrid crops that roughly parallels that thick, blood-red line of forest loss and language loss running straight up from the base of the Sitka Biome. The same Wendy's, Taco Bells, McDonald's, Denny's, Burger Kings, Pizza Huts, and Subways replicate themselves every 50 kilometres or so. The deeper you go down into the Sitka Biome's dead zones, the more the public rest stops take on the appearance of drive-through homeless shelters. Some of them are even fitted out with little soup-kitchen Christian-revival stations.

If you think this is the kind of world Americans really want, then you might be able to convince yourself that they are happy Thomas Jefferson's beloved Taliaferro apple is extinct, and that they don't mind spending $110 billion every year on fast food, up from $6 billion in the 1970s, and that it is of no particular concern to them that their children are the fattest, unhealthiest generation of any in American history. You could say these things are the result of decisions people make, and so that is what people must really want, and no useful inferences can be drawn by comparing the rise of fast food in the United States since 1975 with the 40 percent decline in that country's inflation-adjusted minimum wage over the same period.

But there is no evidence that any of this is what any of us really want. And there is only so much of this kind of thing people can bear.

During the 1990s, a violent reaction erupted over what Greenpeace activists, in their fight against the spread of genetically modified food, had eloquently termed the "deadly banality of capitalism." McDonald's restaurants were bombed, burned, or otherwise destroyed in Athens, Antwerp, Colombia, Copenhagen, London, Rio de Janeiro, St. Petersburg, and, perhaps most famously, Millau, France. It was there that sheep farmer José Bové decided to make a stand against the forces that were turning French cuisine into what he called "McMerde." The Slow Food Movement was born. Bové emerged a folk hero.

In Canada, prairie farmers rallied behind Saskatchewan canola farmer Percy Schmeiser after he was sued by the Monsanto Corporation for patent infringement in 2001. Monsanto's genetically modified, pesticide-resistant canola had "contaminated" Schmeiser's fields to such an extent that it took over most of his crop, a variety he had developed on his own, over decades of trial and error. After a bitter fight through Canada's court system, Schmeiser lost, on the grounds that when he saved the seed from his crop and replanted it, he was planting what he should have known was Monsanto's property. But the case provoked a furor and deepened Canadian farmers' resentment over the increasing control that U.S. agribusiness was exerting in their lives.

Among the wretched of the earth, protests took on especially grim aspects. In India, farmers who had traditionally saved their own seeds from season to season were finding themselves increasingly indebted to the multinationals Cargill and Monsanto as a result of the World Bank's 1998 successes in forcing India to adopt seed-patent strictures favouring the companies. Because the farmers had to return to the companies to buy new hybrid seed every season and to buy the expensive fertilizers and pesticides the new seeds required, they were going broke. They were selling their kidneys to raise the money to keep farming. Outspoken food-security activist Vandana Shiva campaigned to focus world attention on the misery, directly attributing the suicides of 25,000 farmers between 1998 and 2003 to the new seed-patent rules.

In Taiwan, in 2002, about 100,000 farmers stormed the capital, Taipei, demanding an end to government restrictions on the ability of agricultural credit unions to lend money. The farmers were struggling with new rules imposed by the World Trade Organization that opened the country to cheap food-crop imports. Some farmers carried sheaves of wheat. Others carried banners: "If farming dies, the nation dies." The protestors were held back from the presidential offices by barbed-wire barricades and riot police.

In England, a country not known for its terrific food, a wave of revulsion was touched off by the sudden appearance of a McDonald's restaurant at Trafalgar Square. The fight was led by Helen Steel, a part-time bartender, and David Morris, an unemployed single father. McDonald's officials spied on them, sent undercover agents to infiltrate the little group that had sprung up around their campaign, and broke into their offices and stole documents. The McDonald's spies were after evidence to support the corporation's contention that it had been libelled by a Greenpeace leaflet authored by Steel and Morris.

That leaflet alleged, in part, that McDonald's food was bad for you, that the company treated its workers badly, and that tropical rainforests were disappearing to make room for vast Third World cattle ranches supplying the company's voracious demands for beef. The "McLibel" trial was the longest of its kind in British history. Over 20 years after it began, the European Court of Human Rights ruled on February 15, 2005, that the prosecution of Steel and Morris had denied them the right to a fair trial and had unfairly breached their free speech rights. By then, Britons had done such a terrific job of field-trashing by raiding research stations and tearing up genetically modified plants that British food geneticists were leaving the country. Elections were being fought throughout Europe over whether genetically modified foods should be allowed on store shelves.

As is so often the case in these kinds of transformations, nobody really saw it coming.

In 1975, Kent Wheatly started collecting seeds from all those commercially extinct vegetable and fruit varieties that the big seed companies had dropped from their catalogues over the years. There was always a chance that somebody had kept seeds from one of those varieties in an old desk drawer, or that some lost variety still lingered, as a living specimen, in an overgrown apple orchard. As it turned out, there were a lot of drawers and a lot of orchards. Wheatly's Seed Savers Exchange became an international phenomenon. Christine, the cousin

I went for a walk with at the beginning of this book, is a member of the Irish Seed Savers. By 2004, Wheatly and his wife, Diane, were overseeing a complex of offices and seed storage facilities just outside the town of Decorah, Iowa, that housed and displayed a collection of 11,000 distinct varieties of flowers, fruits, grains, herbs, and vegetables.

Throughout Britain, a black market in old fruit and vegetable varieties sprang up in response to Plant Varieties and Seeds Act regulations that forbade the sale of unregistered varieties and demanded fees for registration well beyond what small-scale growers and retailers could afford. The outlaw trade became so popular that it was routinely conducted in the open. Ministry inspectors were often "local boys" who obligingly looked the other way. Bob Flowerdew, an author and BBC radio personality, openly flouted the law. Alan Phillips, the chair of the Brighton and Hove Organic Gardening Group, happily told English journalists, "We're selling seed potatoes for lots of old varieties.... The idea is from Canada but it's catching on. Lots of other places in Britain are looking at it."

One of the sources of the "idea from Canada" was Dan Jason's Seed and Plant Sanctuary for Canada, headquartered on Saltspring Island, just around the curve of Active Pass and across the mouth of Trincomali Channel from the island where I live. Over the years, Saltspring had become a veritable bandit's lair of seed savers, organic farmers, and apple-variety conservers. One Saltspring orchard calls itself Apple Luscious and boasts at least 125 apple varieties. Some of their apples are odd red-fleshed things, and the people who run the orchard don't use any artificial fertilizers—just seaweed, oyster shells, manure, and hay mulch.

Saltspring is also home to a bear of a man whose idea of a good joke is to set his chest hair on fire with a cigarette lighter just to watch how people react. His name is Brian Brett, and my apple-stealing kids could learn a thing or two from him. Brett comes from Cockney-Italian stock and grew up riding his father's souped-up potato-bootlegging truck, racing around Vancouver's backstreets and outwitting the Potato Marketing Board with a degree of cunning sufficient to provide the family with a reasonable income. Along the way, Brett lost his right index finger to an errant shotgun blast. He also writes some of Canada's finest poetry, and he surveys the world from the aerie of his small mixed organic farm in the wooded hills above Cusheon Creek, which commands a grand view over Swanson Channel.

Brett raises pigs, chickens, and sheep, a wide variety of vegetables, and a staggering array of ornamental willows. He is well aware of the northward-creeping sameness moving so inexorably up through the dead zones of the Sitka Biome. Its way had been cleared by the Canada–U.S. Free Trade Agreement, then the North American Free Trade Agreement (NAFTA). Complications arising from NAFTA banned Canadian beef exports to the United States on the grounds that Canadian beef was mad-cow-disease prone and unsafe. The same regulatory regime freed U.S. producers of genetically modified (GM) food from the duty of properly labelling their products, on the grounds that a GM label would stampede the public into the organic-food aisles. In British Columbia, the new regime meant rules to eliminate the "historical differences" between small farmers and corporate farms. Small farmers had always sold their meat and produce directly to consumers; corporate farms rely on giant slaughterhouses governed by the "biosecurity" rules that come with the package of globalized, industrial agriculture.

For farmers like Brian Brett, the restraints on organic farms—ostensibly in aid of preventing mad-cow-disease and avian-flu outbreaks—meant they were no longer allowed to slaughter their own lambs, pigs, cows, and chickens. Instead they were required to haul their animals to the same massive, centralized feedlot-abattoirs that had caused all the "biosecurity" problems in the first place. The food-safety aspects of the new rules had even forced the Katzie tribe, once so famous for its *ska'us* farming, to cancel its annual salmon barbecue.

"The government isn't in cahoots with Monsanto," Brett reckons. "They just think like Monsanto, and we're outside their envelope, so they have to try to crush us."

Brett thought that was just fine. "People will just go underground," he said.

Back home on Mayne Island, Mark Lauckner is unambiguously outside the envelope. From his small glass studio, he makes delightful and eminently marketable *objets d'art* out of glass that he fires in his own kilns from recycled glass scraps. Lauckner is a wiry guy with a beatnik goatee, and when he isn't working in his studio or answering technical-support calls about the glass-recycling furnace plans he offers people for free on his website, he's busy with his bedding-plant business, or otherwise engaged in the solar greenhouse he designed himself, or working in his gardens.

One of Lauckner's proudest passions is his work promoting and selling forgotten, obscure, and heirloom varieties of vegetables. None of them are hybrids. He gets most of his seed from the Wheatlys, the Seed Saver people in Iowa. When I visited Lauckner in his garden one day, he showed me five varieties of garlic, two rare snow peas, four kinds of kale, and a half-dozen types of beans, one of which is called dragon's tongue. It's green with purple stripes, and it shouldn't be confused with the dragon carrot, which is also purple. A pumpkin-like thing was bursting up out of the ground, and Lauckner was quick to point out that it wasn't a pumpkin, and no, it wasn't a squash. It's an oxheart carrot, he said. The W.W. Rawson Company catalogue offered it in 1898, and the John Lewis Childs catalogue offered it in 1900, and then it just disappeared and everybody thought it was extinct. It turned out to be growing in some obscure place under an assumed name.

Lauckner also had four kinds of artichoke, a couple of types of asparagus, 6 kinds of peppers, 12 leaf lettuces, and 3 varieties of beets.

He had Burpee's golden beet, which was known to North American gardeners in 1828, and a white albino beet that doesn't stain, and chioggia beets, which came to North America from Italy in the 1840s and have white and red concentric rings inside them. Of all his ten tomato varieties, not one was red. Tomatoes were rarely red in the days before supermarkets. Lauckner is especially fond of tomatoes.

Among the 2500 baby tomatoes growing in Lauckner's greenhouse was a ground-cherry tomato first described in North Carolina in the 1940s. It's yellow, and rarely bigger than a grape. It grows on a bush that looks like a Chinese lantern plant, and it tastes just like black cherry ice cream. He also had Nebraska wedding tomatoes, which are thick-skinned and bright orange, sweet and juicy, and will last for several weeks, just fine, in your fridge. He had sausage tomatoes, which are speckled green and grow to the size and shape of a cucumber. He had broad ripple tomatoes, known as currant tomatoes because they're smaller than grape tomatoes. The broad ripple was first found growing out of a crack in a sidewalk at the corner of 56th and College streets in Indianapolis, Indiana. Aunt Ruby's German green tomato is a huge, juicy beefsteak of a thing. The fruit is green before it's ripe, and the same colour just afterward, so you have to watch it closely and pick it at just the right shade of green.

Among Mark's regulars are Don and Shanti McDougall, dear friends of mine whose 39-hectare Deaconvale Farm lies on the far side of a deeply wooded hill from my place. Deaconvale is a certified organic farm, and its soil is rich and loamy from the manure of grass-fed cattle and from seaweed hauled up from the island's beaches. The dizzying diversity of the things the McDougalls grow was a big reason for the revival of the weekend farmer's market down at Miners Bay, where I'd seen Tina Farmilo with her big map of the island's old orchards, with all its coloured dots. Deaconvale is where Don and Shanti have chosen to make their stand in life. They are not going away.

Few things are more powerful than humanity's love of the things that grow up out of the ground. Brian Brett on his hill, Mark Lauckner in his garden, Don and Shanti on their farm, Julia Steele at Kew, Tina Farmilo with her orchard map, Alan Phillips at Brighton and Hove—each, in his or her own way, carries on the work extolled in the Apocalypse of St. John. It's the work of strengthening the things that remain. It is honest and useful and necessary work, in the tradition of Princess Augusta, Sir Joseph Banks, Nikolai Ivanovich Vavilov, Pat Mooney, the Wheatlys in Iowa, and those botanists hanging from the cliff faces of Kauai.

None of the people I interviewed during the course of this chapter expect that this necessary and ancient work will end any time soon. Don McDougall, for one, scoffed at the thought.

"You can only push people so far," he said.

A world

The Singing Tree of Chungliyimti

I was a deer hunter, and a boar hunter, and a leopard hunter. When I was a child, sometimes these animals could be killed with stones, and with branches from the trees. It is not that way now.

—Ngowang IV, Angh of Longwa

I'd made no appointment with the Angh of Longwa, but I intended to make his acquaintance regardless. All the roads travelled in the preceding chapters of this book inevitably lead to Longwa, the mother village of a Naga principality in the Patkai Range of the Eastern Himalayas. I'd been told that the angh was perfectly hospitable toward visitors and that I shouldn't be put off by the tales people told about him. One of those stories involved the way he'd handled a delicate diplomatic matter. One of the 42 villages within his realm had refused him its annual tribute of millet and wild cow meat. They say he returned from the errant hamlet with the heads of five insolent noblemen, as gifts for his 13 wives.

I wasn't going to let anything like that worry me now. I'd already made it as far as Khonoma, the ancient mountain citadel of the Angami Nagas. The last 25 kilometres had been an hour-long, vertebra-crushing tilt-a-whirl over a deeply rutted winding road, and we'd arrived long after dark. I'd spent three days down in Assam waiting for my restricted-area permit. It had taken several weeks and the generous help of a retired Indian Army general to negotiate an itinerary that bent the rules without breaking them. The Naga mountains had been closed to the outside world for the most of the previous 130 years. For much of that time, the Naga tribes had been waging a bloody war against the same unravelling of things in the world that this book is about. New Delhi had confirmed truce terms with the main rebel alliance only two years before, ending what had been the longest-running insurgency in Asia. I'd made it to Khonoma. Everything was going to work out fine.

The truce with the Naga rebels was encouraging tourists, but the permit rules were written in such a way as to allow only tightly controlled groups of no fewer than four people, travelling only to approved tourist spots in heavily militarized areas, all the while closely supervised by licensed agents. I wanted to travel alone to places of my own choosing in a jeep with a driver and an interpreter, relying on local guides. The general had said everything had been sorted out, but I still didn't have my restricted-area permit when I arrived in Guwahati, Assam's biggest and grimiest city. I'd had three days to wander around imagining everything that might have got fouled up. But now I wasn't worrying anymore.

When I finally got the permit, it was delivered by no less than Neisatuo Keditsu, the sturdy and smiling president of the Nagaland Tourism Association. It had been issued by the Political Branch of the Home Department, and it had my name on it. It allowed me to enter Nagaland at the border town of Dimapur, 280 kilometres northeast of Guwahati. Neisatuo's job was to get me there, through the military checkpoints, and into Nagaland, hook me up with a jeep and a driver and an interpreter, and see to it that we had local guides when we needed them for the journey on to Longwa, at the top of Tainyai Mountain on the Burmese frontier.

Neisatuo had already made all the arrangements from the complicated itinerary I'd negotiated through Jimmy Singh, the retired general. The security situation, though, was still a bit sticky. I was expected to check in at police barracks in each of the Naga districts we crossed through, but so long as it didn't rain, the roads would be passable all the way through the Patkai Range to Longwa. The permit gave me ten days to get there from Dimapur, and Neisatuo had gotten me up into the mountains and over the switchbacked and potholed roads to Kohima, and then that last rough jag to Khonoma, in a single day's drive. That left a good two days for the visit I'd planned in Khonoma. There'd been a second-storey sleeping room waiting for me there, and a warm blanket for the cold night air.

Khonoma, an old labyrinthine fortress of stone-walled alleys and narrow passageways, was hidden from the outside world until the nineteenth century. The British laid siege to it three times, but they never managed to hold it. Khonoma is also the birthplace of Zapuphizo, the messianic Naga leader who led the Naga resistance against the British in the final years of the Raj and later united the fractious Naga tribes against New Delhi.

Khonoma had been on the front lines of the Naga insurgency from the beginning. But since the truce, its innovations in the work of conserving diversity, rooted in practices from deep in Angami history, were starting to attract international attention. This was why I'd wanted to make a special visit to Khonoma on my way through the mountains to Longwa.

The Patkai Range lies within one of the world's last great redoubts of biological diversity. It's one those places where the United Nations Educational, Scientific and Cultural Organization (UNESCO) finds the "overlapping distribution of ethnolinguistic and biological diversities" in the world, and it has all of those jumbled topographies and wild aggregations of soil types that Nikolai Ivanovich Vavilov identified as preconditions for his epicentres of food-plant diversity. It's one of those haunted crossroads of the biosphere and the ethnosphere, the seen and the unseen, the wild and the tamed.

One of the maps that led me to Nagaland was prepared by Conservation International to show what it calls the world's "hotspots," those bleeding pieces of earth that require special attention because they're so obviously hemorrhaging. To be put on that map, an ecological region has to have at least 1500 endemic plant species, meaning plant species that occur nowhere else in the world. It's like triage: if the place is suffering the loss of at least 70 percent of its original habitat, then it's included. That's because habitat usually means trees, and when you take away the trees, a cascade of extinctions usually follows. At the time that I was writing this book, 25 such places were in that triage ward. One was the Indo-Burman region, which includes the remote mountains where the Nagas live.

Wherever there's a dizzying variety of wild animal species, trees, birds, amphibians, bugs, and flowers, you also tend to find an astonishing diversity in languages, ways of life, fruit varieties, domesticated crop species, and types of livestock. UNESCO calls these places the planet's centres of "megadiversity." Eccentricity of this kind occurs wherever there's a lot of opportunity for give and take between ecosystems, and between people and animals, alongside a lot of opportunity for the opposite, which is isolation—the evolutionary crucible of species, of languages, and of distinct food-plant varieties.

That such special places exist at all is curious, but it's often a simple matter of geography. Where there are mountains and deep river valleys, you get a lot of sharp changes in elevation and a profusion of microclimates, and different kinds of ecosystems just get piled on top of one another. Such places are often mountainous tropical forests, where the growing season is nice and long, there is plentiful rain and fecundity, splendid opportunities for killing and for being killed, and many interesting languages in which to compose poems about it all. In all likelihood, you'll also find an astonishing diversity of beetles. As the British biologist J.B.S. Haldane once observed, "God seems inordinately fond of beetles."

By the beginning of this century, it was almost certain that these centres of megadiversity would be showing up in the cartography of sad farewells. On those maps are such places as the Southern Sierra Madre mountains of Mexico, the northern coasts of Papua New Guinea, the islands of Taiwan, and the Indonesian islands of the Banda Sea. These are also the kinds of places where the languages are dying. In Nigeria, in the highlands of the Bauchi Plateau and throughout the Niger and Benue valleys, more than 80 languages are at risk. By the 1970s, only 30 people could speak Luri. By the 1980s, only 4 old people could speak Holma. Several of these near-dead languages, such as Bukwen, Doka, Dugaza, and Nyam, were spoken only in a single village.

The map that led me directly to the Patkai Range rather than to one of those other places was one the conservation groups Terralingua and the World Wide Fund for Nature prepared for UNESCO. It's a map with an unusual amount of colours, slanted lines, and dots. The colours show the world's biomes—those distinctive communities of habitat types, food plants, animals, and so on. The slanted lines set out small and highly threatened "ecoregions," with more specific types of species assemblages. Black dots mean languages; red dots mean the languages are almost extinct. On that map was a place preposterously crowded with colours and slanted lines but with no dying-language red dots. Instead, an inky profusion of black dots indicated an abundance of small but living languages, especially in that spur of mountains that curves out of the Eastern Himalayas and bends back south and westward again. The Konyak Naga village of Longwa, on Tainyai Mountain, was at its most back-of-beyond corner. Something different was happening up there.

In the region that takes in the Patkai Range, the Indian government officially recognizes 160 tribes that speak at least 30 major languages, but many more smaller languages and hundreds of distinctive dialects can also be heard. Among the Nagas are 31 dialects of Konyak, for instance, and there are 7 Konyak languages among the 20 languages of the Konyak-Bodo-Garo group, which makes up the bulk of the 24-language Jingpho-Bodo-Garo group, part of the 351-member Tibeto-Burman family of languages, which takes up almost all of the 365 Sino-Tibetan languages.

Naga vegetable gardens are just as complicated, with their extraordinarily narrow paths between the wild and the tamed, between natural selection and artificial selection. That's because so many of the cultivated species have such close relatives growing on the nearby mountainsides. Naga gardens grow dozens of endemic varieties of pigeon peas, winged beans, broad beans, and sword beans, and more than 50 varieties of citrus fruits. The jungles in the mountain valleys hold uncounted species of wild banana, so many that the region is widely believed to be the evolutionary cradle of the world's bananas. One brief field study by India's Applied Environmental Research Foundation revealed more than 150 food-plant species growing in the gardens of just three Naga villages. The villagers could rattle off the names of 14 varieties of millet and 32 types of rice.

The entire region tends to fall outside the Western world's field of vision. This is because, on conventional world maps, the Republic of India commonly leaves the general impression of being shaped like an inverted pyramid suspended from the Himalayas, with Bangladesh to the right of it, Pakistan to the left, and the island of Sri Lanka floating in the Indian Ocean off its southern tip. What often goes unnoticed is an area about the size of Britain that lies in the map's upper right-hand corner, appended to the Indian subcontinent by the Siliguri Pass, which in places is only 21 kilometres wide and snakes its way between the top of Bangladesh and the southern borders of Sikkim and Bhutan.

After you emerge from the Siliguri corridor, you keep going across the Plains of Assam, heading roughly east, and eventually you'll find yourself in the Patkai Range. Up there you can stand on a mountain, in the middle of several mountain ranges, and still be within 200 kilometres of several of Asia's greatest rivers, among them the Yangtze, the Chindwin, the Irrawaddy, the Mekong, and the Brahmaputra. Through the ages, the people of the great civilizations of the Indian subcontinent were only dimly aware of the area. The forests at the headwaters of the Brahmaputra River were said to be where Krishna met his bride, and the valleys were said to be filled with the strangest creatures. It's a through-the-looking-glass kind of place.

Among the creatures there is the Mishmi takin, which is an ungulate, *Budorcas taxicolor taxicolor,* a distant cousin of the muskox. Its gold coat was reputedly the source of the golden fleece of Greek myth. Two other rare goat-antelopes live in the mountains, the serow and the goral, and there is also the tahr, a woolly distant cousin of mountain goats. In the valleys, there is the giant one-horned Indian rhinoceros, a solitary behemoth that lives nowhere else on earth. It's the size of a killer whale. An adult male can weigh in at 2200 kilograms.

Exceedingly rare animals inhabit the forests. One of them is the leaf muntjac, the world's smallest deer. It weighs about ten kilograms full-grown and is just half a metre high at the shoulder. The southeast side of the Patkai Range is the only place on earth where you can find one, but you would have to spend a great deal of time looking—the leaf muntjac revealed itself to science only in 1997. Another deer, the brow-antlered sangai, also known as the dancing deer of Manipur, lives only on the floating islands of phumdi grass on Loktak Lake.

There are nine species of non-human primates in the broad-leafed forests, including the slow loris, which looks like a cross between a badger and an accountant, and a tiny tree ape, the long-limbed

Hoolock gibbon, which is about the size of a regular house cat. Through the nineteenth century, rumours surfaced of a "white monkey" known only to certain tribes between the Sankosh and Manas rivers in the hills just northwest of the Brahmaputra River. It was sometimes described as blond, or cream-coloured. Its existence was confirmed only in 1956 when the naturalist Edward Gee found one and gave it the name "golden langur."

There are giant flying squirrels, and even their taxonomical names are beautiful. *Petaurista nobilis* and *Petaurista magnificus* mean "noble rope-dancer" and "magnificent rope-dancer." There are desperately rare red pandas and musk deer in the bamboo forests, and up in the alpine there are blue sheep and snow leopards. Hog deer and barking deer, sloth bears and black bears, tigers and jungle cats, marbled cats and golden cats live among some of the world's rarest birds—Blyth's tragopan, Ward's tragopan, the Derbyan parakeet, immaculate wren babblers, laughing thrushes, and scimitar babblers. Of the five species of hornbills, three are found nowhere else on earth. Forests of alder and pine and juniper and banyan rise up out of 100 species of fleshy mushrooms, scores of endemic rhododendrons, and 600 species of orchid. Sixty indigenous species of bamboo grow there, including a type of giant bamboo that can reach as high as an eight-storey building.

Most of these animals and many of these plant communities are in peril. As those maps of the planet's triage wards made so clear, Nagaland was bleeding badly. But if there was anywhere on earth that stood a chance of surviving the dark and gathering sameness in the world, then this might be the place, I'd reckoned. I'd convinced myself of it. That's why I was huddling against the cold in Khonoma.

That morning, when I had set out with Neisatuo from Guwahati, the sun was rising over the broad Brahmaputra out of a heavy morning mist that lingered over the rice paddies and clung to the bamboo thickets on the hills. Neisatuo drove fast, and we chatted as the villages rolled by. Egrets and doves flew overhead, and flocks of pigs travelled on the road, and soon the country turned to parkland, in the heart of Assam's famous tea estates. It's beautiful country in the springtime, with carefully tended waist-high tea bushes thick upon the ground under shade trees of siran and kala, tidy villages of mud-brick houses, and flower gardens under the palm trees. But when we arrived at Dimapur, it was obvious why this part of the world was at such a precipice, and why extinction loomed so large over everything.

Long before the British came, Dimapur had been the capital of the little kingdom of Kachari, governed by a tribal dynasty that claimed a Naga princess among its ancestors. As recently as the 1970s, Dimapur was a sleepy little town of mysterious stone monuments and ruins. But the Dimapur that Neisatuo and I pulled into was a traffic-choked, raucous little city in the throes of a population explosion. Dimapur's explosive growth was one of the main reasons why Nagaland's overall population has almost tripled, to 1.8 million people, over the previous 20 years.

The population explosion was due mainly to refugees, escaping the collapse of ecological and civil order that was uprooting huge stretches of Bihar, Bangladesh, and Bengal. The massive influx of outsiders was also driving the local Bodos and the Rajbonshis to fight for their own states on the Brahmaputra, and it was the grievance that nurtured Mizo nationalists, who watched helplessly as an exodus of Burmese Chins poured into their homelands. In some parts of the region, the human population was doubling every few years.

At the Nagaland border checkpoint, in the heat of the afternoon, paramilitaries with the Indian government's Assam Rifles stood lazily in their camouflage uniforms, with Indian-made SLR assault rifles slung over their shoulders. Others sat with scarves around their faces and submachine guns in their laps. But half an hour after we'd passed through the checkpoint, we were lurching up the steep canyon of the Chathe River. Dimapur disappeared below us, and everything changed again.

As we coiled through the mountains, passing bamboo-thatched villages and groves of pineapple and genari trees, even the road signs were odd. After Whiskey, Very Risky. Not Safe To Travel At Night. Assam Rifles, Friends Of The Hill People. Safety First, Speed Next. We roared on through the village of Medziphema, and then through Zubza. As dusk fell, women streamed out of forest paths and onto the roads with stacks of firewood on their heads, and chickens raced out of our way.

I called Neisatuo's attention to the It Is Not Rally Enjoy Valley sign, but what followed after that one was Don't Gossip, Let Him Drive. So I spent the time quietly puzzling over Self-Trust Is The Essence Of Heroism until we reached Kohima, 70 kilometres into the mountains above Dimapur. It was long after sunset, and after Kohima we had just that last, grinding, hour-long backroad ascent through the night to Khonoma, the Angami citadel. I was finally through the looking glass.

When I woke up the next morning I could have sworn I was in Machu Picchu.

The morning air was clear and crisp, and Khonoma was coming alive with roosters and songbirds. Young women, shy and smiling, walked quietly along stone-lined passageways. Young men trudged sleepily down narrow lanes on their way out of the village and into the sunrise.

Khonoma sprawls across a sloping mountain ridge about 1500 metres in altitude, commanding a sweeping view of the surrounding countryside. About 3000 people live in Khonoma, and they carry on their lives around the central hearths of little bamboo houses, all jumbled together with the great vaulted tombs of their ancestors. Huge memorial stones rise up out of the ground, commemorating the great feasts and victories of the past. There are pigpens and chicken coops, and flower-garlanded courtyards, and below Khonoma's walls and parapets, meticulously sculpted rice-paddy terraces cascade down the mountain slopes like staircases built by some long-ago race of giants.

The chore of escorting me around and explaining the things I was seeing had fallen to Kevi Meyase, a cheerful 29-year-old part-time carpenter and member of the Army of Khonoma. The women I had seen were heading out into the forest to gather wild bananas, and the sleepy young men were heading off to their ritual bathing places. It was the first day of the Festival of Serenyi, the first of the eight major festivals of the Angami calendar.

Kevi had joined me for a breakfast of boiled eggs and greens at the house I'd been put up in—a guesthouse of sorts on the edge of the village. I was sharing the place with some experts from Bangalore the Indian government had sent to install solar-powered streetlights, an innovation the locals seemed to find endlessly amusing. We left for a walk through the village, followed by a procession of laughing children. When we reached the *rukhuba,* the highest stone tower in Khonoma, Kevi pointed north, down into a valley that fell away in the direction of the Chathe River.

Down there, Kevi said. Our enemies always advance from that way.

The Chathe River empties out of the mountains at Dimapur. That was the way Neisatuo and I had come the night before, on the main road from Assam into Nagaland. It was also from that direction, in the spring of 1850, that a British infantry slowly approached, dragging field artillery up the narrow mountain trails.

For 20 years, Khonoma's warriors had been harrying the British and their lowland allies. It had begun in 1832, when British officers led 700 Manipuri troops on a bloody march through the Angami Nagas' southern mountains. The British objective was to annex a corridor between the Assam plains and the Manipuri lowlands. The soldiers sensibly avoided Khonoma, but through the years Khonoma's warriors had laid down ambuscades at every chance. The British administration in Assam decided that the only sensible thing to do was trounce the ruffians in their secluded mountain stronghold once and for all.

In that first siege of Khonoma, in 1850, mortar fire had little ill effect upon the citadel's stone breastworks and buttresses. The forest paths around the fortress had been sewn with sharp bamboo *panjis* that pierced through the soles of the soldiers' boots, and the narrow ridgetop trails on Khonoma's final approaches were blocked by deep ditches. Arrows, spears, stones, and musket balls rained down on the few soldiers who got anywhere near the walls. The British limped back to the plains.

Nine months later, Major Henry Foquette returned in command of two units of the Assam Light Infantry, a company of militiamen, and a battery of cannons. Ten days into that siege the citadel's defences were breached, and the battle turned to bloody hand-to-hand fighting inside the labyrinth. Most of the defenders managed to creep away through the hidden passageways, covered trenches, and secret trails that led into the mountains.

The lesson the British took from the events of 1850 was that it was probably best to build a fortified stronghold of their own at Kohima, from which heavily armed patrols might occasionally be sent to march around the countryside, waving the Union Jack. The lesson the Khonoma chiefs took from the experience was that they needed better weapons. Over the years, the people drifted back to their beloved citadel. They rebuilt the village and went on with their lives. The warriors, especially the young men of the Merhuma clan, spent their time building up an arsenal of old muzzleloaders and Enfield rifles, obtained in trade from the Kuki tribes in the mountains to the south.

In October 1879, Khonoma's clansmen attacked a British military patrol, leaving 33 dead and 19 wounded. Two weeks later, the Khonoma chiefs led an army of 6000 Nagas in an assault on the British garrison at Kohima. After much bloodshed on both sides, truce talks began, but in the middle of the discussions a 2000-strong column of British infantry and Manipuri riflemen showed up. The Nagas hurried away to their villages. A month later the British attacked Khonoma again. After much more loss of life on both sides, the villagers fled once more into the mountains. Again the citadel was burned, but again the people drifted back over the years, rebuilding everything that had been broken.

Then, on May 16, 1904, a child was born in Khonoma. He was named Zapuphizo, and his birth marked the beginning of the Nagas' encounter with the forces of sameness in the modern world. The Nagas had seen what those forces had done in the world outside—all they had to do was look down from their mountains, to the Plains of Assam.

In the forests of the Naga foothills, a tree once grew that the people of the Singpho tribe held in particular esteem. They used to make a lovely, frothy fermented medicinal beverage from its leaves. In 1823, a Singpho chief presented the Scottish adventurer Robert Bruce with some of the tree's leaves, which ended up at the Royal Botanical Garden at Calcutta. There, the leaves were determined to be from a species of the genus *Camellia,* a relative of the bush the Chinese had for centuries cultivated for their famous green teas that had made so many British merchant mariners fabulously wealthy. The Singpho tribe's wild version was found to produce an exceptionally malty and robust kind of tea.

By the 1850s, British tea plantation owners had taken over great swathes of the Assam plains, where the acidic soils were perfect for cultivating tea. The land was given over to crop monoculture, in huge plantations. By the early twentieth century, India had outstripped China in tea production. And by the end of the century, tea plantations occupied more than 150,000 hectares of Assam's rolling hills and plains. The forests had to give way for all the loveliness of the tea estates, and the Singpho were obliged to make room, too. It was from among the Singpho that the Arunachal Dragon Force guerillas later drew many of its recruits.

Yes, Zapuphizo, Kevi said, looking out over the valley. He was from here, a good man. And up there—Kevi pointed to a mountain peak with a kind of obelisk on top—that is where the women hid with the children, in 1850.

We decided to go for another walk.

Like all Naga villages, Khonoma is divided into *khels*—neighbourhoods occupied by autonomous clans. After we passed Theromo khel, we came to a stone monument, where Kevi stood silently. The monument looked like a giant gravestone. It had been erected in 1995, and on it were the names of 45 men and women of Khonoma "who gave their lives for the vision of a free Naga nation" during the recent insurgency. We passed a monument to Zhudelie Punyu, a famous warrior, and walked through the khel of Kevi's clan, the Merhuma, and then Semo khel.

In Khonoma, each khel is approached through its own stone gate. Within each khel is a clan fort, with its own *morung* and *kwhirheu*. A morung is a longhouse of sorts, the central institution of the young men of the clans. A kwhirheu is a circular gathering place, a small, stone amphitheatre where important meetings are held and trials are conducted. A theft is usually punishable by a fine seven times the value of what was stolen, unless it is a pig, in which case several months' exile is the expected sentence. A murder is punishable by an exile of at least seven years.

We do these things ourselves, Kevi said. We don't have to rely upon India or any other country to solve our problems.

We came to another lookout. There, Kevi said. He swept his hand across the view of a giant staircase rising up out of the deep ravines below and around Khonoma. And there, he said. Now he was pointing up the mountain slopes behind the village. That is where our fields are. The farther up the mountainsides they went, the more the terraces seemed to disappear into virgin forest. But it wasn't virgin forest.

This was the reason I'd come to Khonoma.

It didn't look like it at first, but the rice paddy terraces that rose out of the ravines below and around Khonoma continued on up through the forests and became something else along the way. The farther up the mountainsides the terraces went, the more they seemed to vanish into forest. But the terraces didn't really disappear at all, and it wasn't really forest up there, either. Something else was going on.

This was the frontier between the wild and the tamed, between natural selection and artificial selection, between the wilderness and the domesticated. Around Khonoma, that frontier is not a line or a boundary. It is a netherworld, a deep green canopy of trees reached from the village by an ancient, stone-paved, vine-covered road.

An hour's walk from the village, Kevi and I were making our way along footpaths that led up into tehou trees, which give everything from seeds for necklace beads to planks for coffins. There were teishu trees, skinny-trunked things with a soft pulpy core used for ornaments worn only at certain festivals. Tuzu trees bloom with a beautiful white flower the songbirds are particularly fond of, especially the tsulhe birds, the vudies, and the nunyuno birds. The kevi tree, with edible bark and a pink flower that smells just like tiger balm, is what Kevi's parents had in mind when they named him. But most of the trees are Nepalese alder, *Alnus nepalensis;* they grow the thickest just above the terraces where the rice paddies were giving way to something else.

There were huge tracts of alder stands where it looked as though the branches had all been stripped away. In those places, with the canopy gone, I could see that even on the steep mountainside, well above the rice paddies, the ground under the trees had been sculpted into terraces. In places the terraces were no more than two metres wide, held in place by stone walls. Kevi said it had always been this way. He couldn't remember anyone ever building a terrace.

The fields too have an unseen architecture. Each of Khonoma's khels owns several major tracts, and each tract is divided into small family-owned plots, rarely more than an acre. Each family usually owns no more than four or five plots. But this required more explaining, and Kevi said we would get a better view of things on the other side of the deep valley that falls away below Khonoma. So off we trudged. An hour later, we crossed a stream tumbling out of a little canyon. Something moved beside the trail, between the alders. It was big and black. Then there was another one. And another.

Mithun, Kevi said, smiling.

Few creatures so easily traverse that netherworld between the wild and the tamed as the animal Kevi called a mithun. It confounds description in the languages of environmentalism and conventional taxonomy. Often called a wild buffalo, and closely related to the remnant wild dulong of China's Yunnan province, the mithun is known to taxonomists as *Bos frontalis.* It shares a broad genus with the yak, the domesticated cow, and Europe's extinct auroch. Unlike cows, which graze, mithuns browse, like deer. They are usually black, sometimes brown, sometimes mottled. They are ox-sized things, with massive horns, shy and gentle and lumbering. They once ranged throughout Southeast Asia. By conventional estimates, only 1000 remain in the wild. That, however, depends on the problematic meaning of the word wild.

Look, Kevi said, pointing to the biggest one. In his ear, he said.

It was a little notch—the mark of a Khonoma villager, indicating the mithun's ownership. Village boys keep an eye on the "owned" mithuns that wander the forests, sometimes building fencelike structures to keep them inside certain small valleys or out of cultivated plots and gardens. Sometimes the boys leave salt on a forest trail and blow horns. The mithuns learn to associate salt licks with the sound of the horns, so they're taught, you could say, to come when they're called. Whatever the official estimate of "wild" mithuns, there may be thousands of Naga mithuns living this way. They live like "wild" animals until their owners choose to slaughter them. It's a difficult decision, and the event is important, fraught with ritual obligations and complicated social duties. A male mithun can grow to 1000 kilograms. A lot of people get fed.

We continued on. When we had reached the far side of the valley and come out from a clearing above a cliff, everything started to make more sense.

Far below, where the rice-paddy staircases begin their ascent, thick stands of bamboo follow the stairs in places. The bamboo that climbs up around the hills comes from ten distinct species of various uses to Khonoma, from delectable bamboo shoots to spoons, from baskets to fences, and from rice-drying sheets to construction timber. It also provides the raw material for an intricate network of bamboo-pipe viaducts and stone channels that irrigate the rice-paddy terraces, all joined in such a way that the network orchestrates the descent of water from the mountain streams high above Khonoma.

At different elevations on the staircase, the Khonoma villagers cultivate an astonishing two dozen varieties of rice. The villagers hold no difficult discussions about the varieties' beauty or utility, nor do they grow them to make the case for diversity in food-crop varieties. The people do this because they can, Kevi explained. It is because Nagas have very refined tastes, he said.

As the terraces rise up the mountain slopes, their contents begin to give way to alder, and on our way back across to Khonoma, through those tracts where the limbs had been harvested from the trees, the terraced woodlots were alive with women. The tree limbs grow from trunks that have been subjected to pollarding, which involves major tree surgery and mud plasters. Pollarding is to pruning what sculpture is to etching. The result is an effusion of limbs that grow straight up out of the top of a tree's main trunk. Among these bare trunks the women were busy, carrying alder limbs from place to place on their backs and cutting and sorting the limbs into stacks of even lengths.

All that firewood, and not a single tree had been felled. All that food, but no artificial fertilizers, and no bleeding pieces of earth. That's why Khonoma had been attracting such attention: Khonoma is on the front lines in the world war of extinctions, but it is just going about its business. Villagers are experimenting with different kinds of food crops, to extend the growing season and to participate in the cash economy on their own terms. Their labour is not effortless. It is hard work, and the village council has also banned all wildlife hunting within the orbit of its influence in the mountains. That is especially hard work, given the Nagas' passion for shooting things. The Nagas have always been hunters. In the Patkai Range, it's not uncommon to see men carrying homemade muzzleloading muskets. Guns are ubiquitous in these mountains.

Canada's International Development Research Centre was watching Khonoma, and the Canadian International Development Agency (CIDA) was helping the village bring its example to other Naga tribes. In New Delhi, Khonoma had found support in the government's "green village" initiative. But the most important thing about Khonoma was its distinctive approach to slash-and-burn mountain farming, which didn't result in the kind of deforestation taking such a toll on the forests elsewhere in the northeast and elsewhere in the world's tropical forests—the Sixth Great Extinction's bloody epicentre.

During the 1990s, at least 4000 square kilometres of forest in India's northeast were completely lost as a direct result of slash-and-burn farming. That was just one decade. It was hard to measure the true scale of the loss in the Eastern Himalayas because of that netherworld between virgin forest and everything else. There were secondary and successional forests associated with hundreds of thousands of little *jhum* plots that were planted every year. People had been a part of everything for too long to draw too-clear distinctions. But by the late 1990s, the Forest Survey of India was already taking these ambiguities into account. It could report with some confidence that two-thirds of the mountains of India's northeast were still forested, but that half of those forests were severely degraded. Broad-leaf forests and conifer forests were giving way to bamboo, and bamboo forests were turning into herbaceous weeds. In little Meghalaya, one of Nagaland's neighbours, half the forests had become completely desertified.

By the 1990s, the world was losing millions of hectares of forest every year, almost all in tropical forests—in places just like Nagaland. The immediate causes were often obvious: if you chop all the trees down, the forest won't be there anymore. But the underlying causes could not be so easily described. The forces at work were cultural and demographic.

For years, slash-and-burn farming was dismissed as a threat to the world's forests with about as much significance as John Muir's "birds and squirrels." It's true that large-scale industrial forestry had eclipsed the impacts of slash-and-burn farming, but as with farming everywhere else, the twentieth century changed everything. For centuries, human population levels were low enough throughout the tropics to allow for a slash-and-burn approach to farming that was perfectly sustainable. But in most of the tropical world, a corner was turned somewhere around the middle of the twentieth century, when the planet's human population doubled.

The crumbling of the old order was particularly obvious in the Patkai Range. Naga economies had been based on slash-and-burn crop rotation for centuries. By the 1980s, though, the tribal population was growing by 4 percent every year. It might not sound like much, but at that rate, a population doubles every 20 years or so.

Slash-and-burn farming is now a major contributor to the global calamity of tropical forest loss. And when you add illegal logging, wildfires, and climate change to the picture, it gets even nastier. In 1997 and 1998, a killing smog covered a vast area of Southeast Asia, stretching from the Philippines and Malaysia to Thailand. The forests of Indonesia were on fire. More than two million hectares of forest went up in smoke, mainly because of runaway slash-and-burn operations. Schools and airports were closed. Hundreds of people died. In 2000, just in the southern half of the African continent, roughly 200 million hectares of forests burned to cinders.

Most of the world's forest loss is occurring in the nation-states of Argentina, Brazil, the Democratic Republic of Congo, Indonesia, Mexico, and Burma. But in the mountains of Nagaland and the neighbouring states of Mizoram, Arunachal Pradesh, and Manipur the losses have been as bad as anywhere.

If Khonoma has a contribution to make to the conversation, it is not alone. Some of the Angamis' neighbours over the mountains in Burma have long practised a similar sort of pollarding, with the same species. About 400 kilometres east, the Lisu, Dai, and Hwui tribes of China's Yunnan province have been doing much the same kind of thing for longer than anyone can remember, using the same kind of alder as the Angamis at Khonoma. The practice also occurs in Northern Luzon in the Philippines, where villagers use a tree from the same genus but a different species, *Alnus japonica.* In Uganda's highlands, villagers use *Alnus acuminita.*

But what Khonoma is doing is particularly impressive.

During that walk with Kevi in the terraces above Khonoma, small, controlled fires smouldered on the ground where the forest canopy had been chopped away, leaving the terraces rich and black with nitrogen and carbon. Where the ground had been burned a year or two before, mustard and Job's tears and tea and millet were already growing up in the shade of new alder limbs. On the narrow terraces, where the alder limbs were thick and heavy, there were yams, chilies, pumpkins, garlic, and cardamom. Once the canopy covered everything again, the terraces would be left fallow and the alder would continue its work of replenishing the soil's nutrients. A couple of years later, the women would be back again to burn the underbrush and cut and stack their firewood, and it starts all over again. It might begin with a bit of maize, or one of those innumerable bean species the Angamis like so much. It might begin with a certain wild species of tomato that grows only in these hills but isn't really "wild" at all—it just doesn't end up in the markets, because the Khonoma people have only so much of it and they prefer to keep it for themselves.

A wild apple tree grows here that the villagers have recently taken to cultivating. I took an apple from one of those trees. It was tart and pithy, and I'd never tasted anything like it. But the experience was as melancholy as the taste of the apples my children had been stealing from that overgrown orchard back home.

On my last night in Khonoma, I ate a dog. Well, some of a dog, at least. It was a bit spicy, but otherwise not bad. Nagas like dog meat, and they love spices, and they like company around the fire. Neisatuo had driven in from Kohima for a visit. We all had a hearty meal and told stories late into the night.

I left Khonoma for the long journey to Longwa in a hopeful mood. My new companions were Visevor Nagi, at the wheel of a sturdy Tata Sumo jeep, and Khrienuo Kense, a bright young woman from the Angami village of Tuophema, who was my interpreter. Khrienuo was fluent in English and Angami and she spoke some Hindi, but most importantly she was fluent in Nagamese, the creole that allows the Nagas to talk with one another across their language boundaries. There are many boundaries.

On the very first day out from Konoma, we travelled through the territories of the Lhota Nagas, and then we crossed the Doyang River into the mountains of the Sema Nagas. As the days passed and we crossed from one tribal territory into the next, and from one language territory into the next, the borders were marked by abrupt changes in the design patterns on the shawls and capes worn by the women, troops of them, making their way to and from their small crop fields in the forest. Sometimes the crossing of a frontier would be signalled by a dramatic variation in the style of the villages' bamboo houses.

The roads, such as they were, followed the mountain ridges, because Naga villages are always on hilltops, ridges, peaks, or mountain spines—tigers rarely attack uphill, they say, and hilltops were easier to defend, in the old days, against enemies. Most Naga villages are still beyond the reach of roads, but now and again we'd come across Bihari road-building gangs. They were entire Bihari villages living in temporary roadside tin-roof shacks, breaking stones with their bare hands, doing work the Naga tribes would not stoop to do, at wages in rupees amounting to less than $2 U.S. a day. Every few weeks a Bihari camp would pick up and move along as their road punched farther along, connecting all those Naga villages to Kohima and Guwahati and the world outside.

So up and down we went, slowly winding our way to Longwa, at an average speed of just 25 kilometres an hour. An encounter with another vehicle was a rare thing, and it was usually a military vehicle. We'd come upon an army checkpoint every few hours—a little tent with an army truck and an officiously polite Assam Rifles officer to check our papers and make a cursory search of our jeep.

As one mountain range fell away into the next, the degenerating economy of slash-and-burn farming erupted into clouds of smoke and ash that billowed down the valleys. Whenever we approached anything that even resembled a town, the same denuded landscapes appeared, and flames licked their way up the mountainsides all around us. There were days when it was as though the entire world was on fire.

The immediate causes are fairly easy to figure out. As the Naga population expanded, their slash-and-burn cycles, known as jhum rotations, contracted. Modern medicine had sharply reduced the infant mortality rate. People were living longer. Where jhum fields were once left to lie fallow for 30 years, farmers were returning to burn and plant far more frequently, sometimes after only five years. In Khonoma, where villagers terraced their jhum plots and relied heavily on the stable root systems and nitrogen-fixing properties of alder, you can get away with cycles that short. But if you keep cutting the trees down and setting everything on fire, things start unravelling. The forest disappears, the soil is rapidly depleted of its nutrients, everything blows away in smoke and dust and ash, and all you have left is weeds.

In the great vortex of extinction, there are always those cycles within cycles. There are ecological forces, cultural forces, and demographic forces.

A collapse of order was driving refugees out of Bihar and Bangladesh into Dimapur, but the Naga tribal population was rising as well, and not just because of better medicines. It was because the Nagas are Baptists. Or at least they're avowedly so, having laid down their Baptist faith upon an ancient foundation of animist practices. The Naga conversion was set off by the brief forays of nineteenth-century American evangelists. The Indian government closed the mountains to missionaries in the 1950s, but the Nagas have been sculpting, terracing, and pollarding their version of Christianity ever since.

When the Nagas became Baptists, old population-growth restraints gave way. The main restraints that allowed Naga jhum rotations to spin so slowly and sustainably for so long are not easily put in the category of neolithic ecological wisdom. The Nagas practised ritual headhunting, and their leisure time was often given over to intertribal warfare. Before they were Baptists, the Nagas believed that the productivity of their jhum fields had to be replenished by the severed heads of their enemies. These kinds of practices can be pretty effective at population control.

These were also the reasons the British were so wary of the Nagas, who were generally disinclined to back down in their many standoffs with British regiments. By the 1870s, the British decided to cordon them off behind a huge no-go area. Later, after the Nagas converted, they weren't any less fright-making. By becoming Baptists, the tribes became more or less united. And later still, after the British left India, the Nagas weren't any less contemptuous of the government in New Delhi. They demanded independence. The long and bloody Naga insurgency began.

To the outside world, the Naga revolutionaries didn't fit neatly into any of the usual categories. They could not be counted among all those other Third World movements that sprang up against the old colonial powers: the Nagas were fighting India, which had itself only just won independence from one of those powers. Zapuphizo, that fiery Angami who came to lead the insurgency, began his seditious activities while working as an insurance salesman with the Calcutta branch of the Sun Life Assurance Company of Canada. You couldn't put the Naga insurgency into the category of those proxy wars the United States and the Soviet Union were orchestrating around the world, either. There were no "green revolution" remedies. Rebels were moving through the mountains, and ambushes, attacks, counter-attacks, and massacres were taking place, but Nagas didn't issue Marxist communiqués with demands to control the means of production. They didn't even seem to be demanding an independent country, exactly. The principal article of faith the united Naga tribes had carefully crafted to articulate their demands stated, "Every Naga village is a republic in its own right."

New Delhi's response was a combination of bloody repression and isolation. India's soldiers behaved badly during their occasional occupations of Naga villages; atrocities were committed. India revived the British no-go strategy, imposing "inner line" permits to keep Indian citizens away. Foreigners were barred as a general rule, allowed in only for approved official-business visits and usually confined to the few main towns. Even the permit that allowed me to travel through the Patkai Range to Longwa was issued under the Foreigners Protected Areas Order of 1958, which was a throwback to the Excluded Areas law of 1935, which was a formalization of the British isolation strategy of 1876. During the bloodiest moments of the insurgency, novelist Graham Greene used to write letters to *The Times* protesting India's policy of barring foreign journalists from the mountains. But most people couldn't even find the place on a map.

The Nagas had always been difficult to situate, in any kind of category. Nobody could even say where the name "Naga" came from or even what it meant: some say it comes from a Hindi word for serpent, others that it originated in a Burmese word for perforated ears, since so many Nagas are fond of gigantic earrings. At least 18 major separately named tribal groupings identify as Nagas. The main tribes are the Angami, Ao, Konyak, Sangtam, Rengma, Phom, Chang, and Chekasang. Some are autocratic, like the Konyak. Others are strictly republican, like the Angami. Some tribes look vaguely Mongolian. Some resemble Vietnamese Hmongs. There is no anthropological consensus about where any of them came from.

Writing in 1915, J.H. Hutton, whose weighty monograph on the Angami Nagas remains a standard ethnographic text, admitted to being amused by the wild speculation about the origins of the northeast tribes. He jokingly suggested there might come a day when some wag would posit the theory that they had descended from one of the ten lost tribes of Israel. In the 1990s, Jerusalem rabbi Eliyahu Avichail said he had identified 3500 Mizo tribesmen who claimed just that. By 2003, more than 800 Mizo people had settled in Israel.

As India consolidated itself out of the British Empire, constituting a republic with an amalgam of states, provinces, and territories with fixed boundaries, it had a particularly difficult time in the complicated terrain of that vast territory on the other side of the Siliguri Pass. In 1962, India and China waged a pointless and thankfully brief war, ostensibly about where their borders should be in the Eastern Himalayas, and the entire northeast region ended up a high-paranoia security zone. The border with China was never resolved, and it wasn't until the 1980s that India's border with Burma was fully agreed upon and surveyed. By that time, India was starting to grant local government powers in the region. The Nagas fared badly in all this.

Half the Naga country ended up in Burma's Kachin state and Sagaing territory. Then half of what was left on the Indian side of the border was parcelled out in bits to the seven states India was conjuring up out of the old North-East Frontier Agency and the far-flung districts the British had administered from Assam—states that only vaguely reflected the linguistic and ethnic realities on the ground. The state of Nagaland ended up taking in the mountains between Manipur in the south and Arunachal Pradesh in the northeast, lying astride the Patkai Range and several smaller ranges.

The Nagas were by no means alone in their grievances. By the 1990s at least 30 armed groups operated in the region. The National Democratic Front of Bodoland (NDFB) wanted a sovereign Bodoland established somewhere on the north bank of the Brahmaputra. The separatist Kamatapur Liberation Organisation (KLO) made hit-and-run attacks on government troops from bases hidden in the mountains of Bhutan. In Arunachal Pradesh, separatists with the Arunachal Dragon Force (ADF), composed mainly of young men from the Khamti, Singpho, and Tangsa tribes, were said to be setting down roots among their cousins across the border in China. To the southwest, the Bru National Liberation Front (BNLF) was fighting for a semi-sovereign state for the Reang tribespeople in Mizoram. In the nearby mountains, the Kuki Liberation Army (KLA) busied itself kidnapping people in Manipur.

A few months before I arrived in Guwahati, a Meghalaya state minister was arrested on charges of "abetting militancy" by harbouring members of the Achik National Volunteers (ANV). The minister's name was Adolph Lu Hitler Marak. One wonders what his mother was thinking. Once you had all the names sorted out, they changed again. A few weeks after Marak's arrest, the All Tripura Tribal Force (ATTF), which was the armed wing of the Tripura Peoples' Democratic Front (TPDF), changed its name to the Revolutionary Peoples' Army (RPA). A few weeks later, the banned National Liberation Front of Tripura (NLFT) changed its name to simply "Plungers and Rangers," perhaps to distinguish itself from all the other acronyms.

Out of all of this confusion, the Nagas could look back on 130 years of hardship and profound cultural change and upheaval, and they could still say that they had held on. They'd kept their heartland and waged a valiant resistance. But thousands of lives were lost, and in the end the struggle degenerated into fratricidal, interfactional warfare. By the time the Naga tribal leadership settled for peace with New Delhi, the world had changed utterly: the jhum rotation cycles were contracting; the forests and all their animals and birds were disappearing. The Nagas' governing tribal council saw in the 2003 truce terms a historic opportunity to put aside the old muskets and foot-piercing panjis and build new kinds of barricades and battlements. Villagers were already streaming to the parapets Khonoma was defending.

Between 1995 and 1997, more than 800 Naga villages planted more than five million trees in Khonoma-style jhum plots throughout Nagaland. More than 100,000 of these trees were from several indigenous species that forest scientists couldn't even identify. Some villages were using alder, some were using gomari trees, and some preferred hollock. Some were planting them in the Khonoma style; others were devising their own variations on Khonoma's narrow terraces. In 1999, Nagaland's state government delivered more than 10 million seedlings to Naga villages, which was no mean feat, since roads hadn't yet reached most of them. The result of all this was that 32,000 hectares of Nagaland that would otherwise have been largely barren were covered in trees.

As we moved through the mountains to Longwa, there were days when the jungle valleys were lush and alive, riven with creeks and rivers of swift and cold water, and the forests lay heavy and green on the mountains. And always, in the middle of nowhere, the strange road signs. Give Beer To Those Who Are Perishing Of Thirst. We Need More Passion Than Fashion. Start Early, Drive Slowly, Reach Safely.

Quiet Please, No Politics. The boundaries between tribal territories announced themselves in older sorts of signs. Here, the men carry only a *dao,* the traditional Naga sword; this is Phom territory. There, the men carry their daos in sheaths on their backs; this is Sangtem territory. Now, all the men carry muskets, and their capes are differently coloured; this must be Konyak country.

Even when the sky was clear and without a mote of ash or dust and the mountains were green and hawks hovered above the deep valleys, and even in the most remote villages, the outside world was moving in and rearranging things. In some villages, like Longkum, it was happening in small and subtle ways.

Longkhum is an Ao tribe village of about 2500 people at the top of a mountain in the Ongbangkong Range. After death, the people used to say, the souls of all Ao people return to Longkhum. The village is at the end of a rutted and winding road that switches back up through jungle and forest to an altitude of 1800 metres. You know Longkhum is an Ao village because in the older sections it's a bit like a big treehouse, with houses built on bamboo stilts. A lot of village life still unfolds on rickety verandahs, and on one of those verandahs, in Chongli khel, an 83-year-old man named Tingmapa sat by himself, stripping bamboo to make a large mat of the kind used in Longkhum for drying rice greens. Tingmapa's 72-year-old wife, Sayamenla, was down in the forest collecting firewood.

Tingmapa was the last animist of his khel, which was otherwise Baptist, and partly Christian Revivalist. He knew only a few fellow animists, among the old people of the Yemrong and Sangpang khels. By the new faiths, the souls of the dead took a somewhat different path than the one Tingmapa expected them to take. But it wasn't that. Christianity is fine, Tingmapa said. It wasn't the roads, or the radio, or even the television. A few people have television sets in Longkhum, but the electricity is on sometimes and off sometimes, and there is nothing on television in the Ao language, anyway, Tingmapa explained.

He continued his work with his dao, deftly cutting narrow strips of bamboo into even narrower strips, and from those thin strips he peeled away the inner bark from the outer bark, and the pile at his side grew. It takes five days to make a proper bamboo sheet for drying rice greens: one day to gather the bamboo in the forest, another day to cut and strip and peel the bamboo, and three days to weave it all together into a mat.

One day, someone in the village spread a basket of rice greens on a sheet of blue plastic tarpaulin, laid out in the sun. Then more people started drying their rice that way. It all just seemed such a modern and forward-thinking thing to do. Everyone in Longkhum seemed to still agree that Tingmapa's mats were superior, but the blue plastic tarpaulins were finding favour just the same.

In other villages, like Tuensang, the outside world was rearranging things with rather more cunning.

Tuensang, a village of 10,000 people on a ridge overlooking the Palak Valley, is the largest of the 56 villages of the Chang tribe, and its beautiful houses were long a matter of great pride among the people who lived there. The Kungsho, Lomo, Pilashi, and Chongbo clans were happy to boast of their broad, sweeping roofs thatched with the palm of the lau tree and their gables hanging almost to the ground. The forepeaks of these roofs, like the prows of great ships, sheltered cozy porches around the front doors. There are still such houses in Tuensang. Some houses are elaborately decorated with wild boar skulls and dog skulls around their front doors, and in the more well-to-do parts of the village the entranceways are adorned with mithun skulls, elephant skulls, and antelope skulls.

At one such traditional house, modest but nevertheless roofed in the old style with lau palm thatches, the doorway was decorated only by a long necklace of eggshells, to bring good luck to the 20 chickens in a tiny, fenced courtyard. Chongsen, a 44-year-old Chongbo clan leader, sat at a hearth fire on an immaculately swept dirt floor in the middle of the house he shared with his wife, Mokjing, their five children, his younger sister, and a cousin. With his three-year-old daughter, Junglumla, on his lap, Chongsen tried to explain how it had come to pass that in the space of 20 years, four out of every five houses in Tuensang had given away their old beauty, and their roofs had been covered with rusty, riveted pieces of tin.

A lightbulb hanging from a wire suspended from the meat-drying rack above the hearth flickered for a moment, and then the light was gone again. Everything is changing, Chongsen said. They want us to put tin roofs on our houses so we will not appear to be backward, so we will not be embarrassed when visitors come to see us.

It was a complicated story involving subsidies New Delhi provided to the Nagaland state government, which were in turn passed on to householders in Chang villages like Tuensang, so long as they agreed to do away with their grand and sweeping lau-thatched roofs. Government inspectors routinely made the rounds of the village to see that everything was just so. The new tin roofs were said to be able to last 50 years, but they were a recent innovation so no one knew whether this was true. The old roofs, if made properly and in accordance with a heavily ritualized process that engaged at least 20 labourers over three days, would last perhaps 40 years, Chongsen said.

As a clan leader, it was part of Chongsen's job to see that the tin roofs were replacing the old ones according to various schedules in the state's five-year plans. He didn't like it, but he had to do it. Chongsen said he wouldn't think of making his own roof out of sheets of tin. Tuensang's clan councils also lamented the loss of the village's beautiful roofs. Their plans for a small tourist village were amended to ensure that all the guesthouses would have lovely lau-thatched roofs, in the old style, with forepeaks like the prows of great ships.

At Langmeang, there were no plans for a tourist village. There was a problem.

Langmeang is a remote Konyak village of about 1200 in the Chompang Mountains, at the end of a deeply rutted track punched through in the 1980s, and the arrival of a jeep with some faraway Nagas and a white man was a pretty big event. As a red sun sank behind the mountains in the west, we were immediately surrounded by about 150 laughing and cheering children who escorted us up to the highest ridge in the village, where the main morung stood in the shade of a remnant copse of yunchuk trees and lok trees. In the morung courtyard, an informal gathering of the village elders was underway. The old men sat quietly, smiles showing through the broad zigzag tattoos on their faces, around an open fire, beside a large obelisk and a stone altar. It was as though we had been expected.

The 32-year-old Angh of Langmeang, Dujai, happened to be seated with the old men. We exchanged greetings through my translator, Khrienuo, in Nagamese. Straightaway, Dujai said something to a little boy, who ran off and returned a minute later with a flashlight. Then the boy and two of his companions gestured that I should go inside the morung. A cheer went up among the old men.

The dark entranceway into the longhouse was decorated with mithun skulls and stag antlers. Animal shapes were carved deeply into the roof beams and the morung's huge support posts, but I couldn't make out what they were. The boys led me into a small chamber, a sort of anteroom. In the pitch dark behind a bamboo screen, the flashlight beam illuminated an open sarcophagus filled with human skulls.

The boys picked three skulls from the heap and brought them into the main chamber of the morung, where they laid them on the floor. One of the skulls had been painted with the tattoo design I'd seen on one of the faces of the old men outside. Some of the men came inside and looked at the skulls in the flashlight beam, and the boys giggled to themselves. When I joined the others outside, more cheers arose.

Langmeang was the last village in the area to have accepted the Christian faith, around the same time the road came. Now, the village desperately wanted tourists, Dujai explained, but the village elders felt that the only thing Langmeang had to offer was the grisly spectacle of those heads. In neighbouring villages, where everyone had become Baptists so many years before, all the skull reliquaries had been buried or burned.

Khrienuo listened carefully to Dujai, who spoke slowly. She repeated his sentences one by one.

We want to be like you, in the developed countries, Dujai said. We want to be able to send our children to good schools. Learning English is good, too. The animal skulls you see everywhere, and the bear paws, and the tiger teeth. These are from animals that are not here anymore. So we want to have tourists come. But there is a problem.

Khrienuo and Dujai exchanged several sentences on their own in Nagamese, and then Khrienuo explained the problem of the skulls.

On the one hand, Langmeang's villagers were afraid that the world would think them backward and savage. On the other, the state government was encouraging them to display the skulls, not bury them as the other villages had done, because tourists would come to see them. But the skulls had been taken over many generations in attacks on the people of nearby Totok village, and in attacks on the villages of Chunyu, Chen, and Chinkao. All these villages once had their own reliquaries, full of Langmeang skulls. In those other villages, the relatives of the people whose skulls Langmeang still held had heard that a German tourist had been shown the skulls, and then other tourists had come to see them. This was upsetting to the people of Totok and Chunyu and Chen and Chinkao. Some of them warned quietly that the spirits of those dead people would bring ill luck upon Langmeang if their skulls were ever brought out into the sun.

So a compromise of sorts was struck. For the time being, visitors would be brought into the morung, and they would be shown the skulls in the dark, but not in the light of day.

Dujai then spoke forcefully to Khrienuo. The old men around him listened quietly.

He wants me to say, Khrienuo explained, that his children should have what your children have, in the developed countries. He said that this is why they want tourists to come and visit them here in Langmeang.

Khrienuo listened carefully to Dujai, and continued: He says, we want to be part of the outside world. So we will show the skulls, and then we will see if anything bad happens.

It was a different story altogether in Ungma.

In this, the oldest and largest village of the Ao tribe, about 5000 people from the Pongener, Longkumer, and Jamer clans live in the Ao tree-house style, their houses built on pilings driven into steep cliffs that fall away into the clouds from a series of hilltops in the Mokukchung Mountains overlooking the Dikhu River Valley. Compared with most Naga villages, Ungma appears cosmopolitan, even worldly, which is to say it boasts about six blocks of pavement, and its huge Baptist church, built in 1912, seats at least a thousand. But with all its little houses built on sturdy bamboo stilts, and its pigpens and chicken coops, the village was still poised at the edge of the neolithic.

Ungma remained a community of subsistence farmers whose houses were decorated with animal skulls, and who rose at dawn to make their tea at open-fire hearths, and who went about their lives carrying their daos and their baskets on their way to the fields. They still stopped on footpaths along the way to chat with their neighbours in the musical Chongli dialect, and they burned, planted, and reaped from their little hillside plots as their ancestors had. But within the space of 20 years, Ungma had also become a place connected to the outside world by roads, and then by telephones, and then by 60 television stations, at least in those few homes that had television sets and could afford the 150 rupees to splice into the cable from the village satellite dish on those rare days when the electricity was working.

Then a three-storey conical structure was built on one of Ungma's hilltops to serve as a guesthouse for the tourists who were beginning to come to Nagaland. From a distance, it resembled some strange medieval lighthouse, and it looked like one inside, too, with a winding staircase from the ground floor to the roof. I was the first foreign guest to stay there, so I was greeted by various village officials the morning I arrived in Ungma, and they took me around to shake people's hands. The next day I was brought to an assembly in the village square for a performance of raucous and colourful mock-warfare dances. Young barefoot men carrying long Ao spears and shields whirled and hooted in a riot of colour, in hornbill-feather headdresses, red-dyed tree-bark breastplates, boar-tusk necklaces, elephant's trunk armlets, and black kilts that jingled with cowrie shells. It was all very grand.

Much of this elaborate hospitality had been arranged and orchestrated by Sangyusang Pongener, a beloved Naga folklorist, singer, and composer. Famous well beyond the Ao tribe, wiry, exuberant, and fiercely attentive to the Aos' strict rules of hospitality, the 55-year-old father of five invited me over to his house one night. He'd gathered several of the singers of Ungma around his hearth, and we stayed up late, drinking rice wine made from the mapok paddy, and discussing the affairs of the world, telling stories, and singing songs.

Tourism will be good, Sangyusang said. It is actually reminding the young people who they are. They are dancing again, and they are wearing the old regalia, he said. Farming isn't enough anymore. There is only so much land, and there are many people.

In the 1920s, a British census reckoned that there were 30,599 Ao people scattered in the several dozen villages perched on hilltops between the Dikhu River to the west and the Disai River to the east.

By the end of the 1990s, the population had risen to 141,000, more than a fourfold increase.

Tourism will have to be part of some new economy, Sangyusang continued, and it is good for the young people. Too many of the children were watching television, and they were not learning the old stories and songs. Because the Naga people have no written history, these things were disappearing quickly. But with tourism, he said, the young people are beginning to learn about their culture again. They see that it is important.

But now we must sing another song.

Are you bathing now? What other birds will come and visit me, apart from you?

And so it went like this, deep into the night—old men wandering in and out around the fire, Sangyusang leading the songs, and sometimes his wife, Rongsenpenla, leading the songs. They were ballads, lively rounds, love songs, and laments.

You are like a bird in a king's house. What makes you test me so?

The songs were mostly sung a cappella, although sometimes Sangyusang accompanied a tune on the *kongki,* a single-stringed violin made from a gourd and a piece of bamboo, with a horsetail hair for a string.

Ao songs are sung in three voices, Sangyusang explained. There is the father voice, the mother voice, and the child voice. Most Ao songs have three parts. One part is sung by the *tetenyur,* who tells the song's story, usually in the mother voice, and sings the first lines of each stanza. The tetenyur is answered by the *tenerok,* whose response is usually sung in the child voice. These voices are accompanied by the *tazungsemers,* whose chorus is normally sung in the father voice.

Yes, let us not go into the fields this morning.

The songs are almost always composed for one tetenyur, one tenerok, and as many tazungsemers as might be gathered around a hearth fire. Sangyusang almost always sang the part of the tetenyur.

Between songs there was much discussion about whose songs were the best among the Nagas and whose voices were the best. Among the Angamis and Aos at Sangyusang's house that night, everyone agreed that the Chekasang tribe had the best songs and were the best singers. Still, in the Ao tradition, if a song did not sound quite right, there were many other ways to sing it.

Maybe some new bird has come to the place where you sleep, and this has made you forget me.

Every Ao song has 12 versions, Sangyusang explained: When you first learn the song. When you sit in a house. When you are sitting just outside your house. When you sing it to your sweetheart. When you are alone in the fields. When you are working in the fields. When you are lonely for your sweetheart. When you are on your way into the jungle. When you are pounding rice. When your arms are around your friends. When you're dancing. When you are accompanied by the one-stringed violin, the kongki.

You are like flowers in front of a king's palace.

And there is another thing, Sangyusang said. We sing our old songs in a language that is used only for singing.

How this language came about is accounted for in an epic tale from the beginning of Ao history, at the place where the first Ao people emerged from under the ground. Obelisks still mark that place in the forest, at Lungterok, on the far side of the Dikhu River, on Chungliyimti Mountain, the big mountain I could see in the distance from the roof of the lighthouse-like guesthouse.

Every morning, a beautiful young woman of the Jamer clan went to bathe in a pond in the forest, and she would hang her black shawl from the limbs of a sangwa tree that hung over the pond.

The shawl you wear is the most beautiful of any in Chungliyimti.

One morning, as the young woman was getting out of the pond, she reached for her shawl, and the limbs of the tree embraced her, and the tree began to sing a song.

You are the most beautiful of the women of Chungliyimti. You are the most beautiful orchid in all of Chungliyimti.

Every morning, the girl went to her bathing pond, and every morning the tree embraced her and sang to her. The boys of the village became aware that this was happening, and they watched, and listened to the tree's songs. Eventually, the girl's parents discovered these things. Her father flew into a rage. He ordered his daughter to stay inside their house, and he told the boys of the village to cut down the tree.

Please do not be sad anymore. I will change.

As the girl watched from a hole in the thatched roof of her house, a splinter flew from the tree during its descent and killed her. Her parents mourned, and the boys of the village mourned, and to this day, out of modesty, the women of Ungma will not sleep on a bench made from the wood of the sangwa tree. And the Ao people sing their old songs only in the language they once learned from the singing tree of Chungliyimti, in the days when the land was young and the world was new.

Two days later, spattered in grime and muck, we stood beside the Tata Sumo, its wheels spinning in the mud. Bilong Nagi, a Konyak guide who'd joined us a couple of days earlier, waved his arms for Visevor, our driver, to stop.

We will not be able to make it to Longwa, Bilong said.

We'd left the crowded little market town of Mon almost two hours before but had covered only about ten kilometres. Longwa was still 32 kilometres away.

We'll try again, I said.

Bilong and I put our shoulders to the back of the Sumo and pushed as hard as we could. Khrienuo stood and watched, sensibly avoiding the folly of trying to force a massive two-wheel-drive vehicle with slick tires up a wet and muddy hill. It had rained the night before, and all over Nagaland the roads were gumbo. But we kept trying anyway. We pushed and groaned and the wheels spun until we were mud-spattered and spitting. Then we turned back.

On the return trip to Mon, I kept seeing ghosts on the road. The unicorn-horned Elasmotherium that Ahmad ibn Fadlan claimed he saw in the forests of the Volga River. The Carolina parakeet. The Sicilian pony elephant. The last Irish wolf, 1786. The last Pyrenean ibex, January 6, 2000. Vancouver's last pair of crested mynahs, run over by cars, February 2003. That Hawaiian honeycreeper, the po`ouli, in its cage, November 26, 2004. It was like that all the way back to Mon.

I'd come to the Patkai Range because it was a little world unto itself, and it happened to be caught in the headlights at one of those haunted crossroads of the earth, where the squiggly lines and colours and dots showed up on all those maps. I'd convinced myself that the mountains up here stood a good chance of surviving the dark and gathering sameness in the world. I spent a lot of time thinking about that on the way back to Mon. I spent a lot of time thinking about it a few months later when at least two dozen people were killed in a bomb blast in Dimapur. I am still convinced.

The famous dancing deer of Manipur had been reduced to a single, barely viable population of 150 animals on islands of phumdi grass that were thinning and shrinking every year because of the operations of a 103-megawatt dam built at the outlet of Loktak Lake in 1983. During the 1990s, Edward Gee's "white monkey," the golden langur, lost one-third of the only forests in which it has ever lived. The langurs had become trapped in isolated wooded pockets on the north bank of the Brahmaputra, reduced to the state of village beggars, stealing fruit from orchards. The same sort of thing was happening to the Konyak tribesmen who had once been such proud hunters of boar and deer. Though the men still wouldn't so much as go for a short walk without taking their beautiful homemade muskets along, they had been reduced to shooting little songbirds out of the trees.

But from the mountains near Longwa, you can see as far as the Sangpang Range, and those mountains lie within a vast area that the Burmese government is planning to set aside as a tiger preserve—an area larger than the entire state of Nagaland. Seventy percent of the forests of Arunachal Pradesh, only a couple of days' hard hiking from Mon, are still intact. The golden-fleeced Mishmi takin holds its own there. The great one-horned rhinoceros that had been reduced to the marsh grass and mud wallows of the Easter Himalayan foothills had dwindled to about 300 animals by the end of the nineteenth century, but it survived all the ravages of the twentieth century, and its numbers are rising again. Most Naga children still speak their own languages. More than a century and a half after troops of infantry dragged their field artillery up the trails from the Chathe River Valley, the old Angami citadel still stands, and hundreds of Naga villages are planting trees and pollarding trees like in Khonoma, in that netherword between the wild and the tamed.

But the main reason I'm convinced there is still cause to hope for those mountains at the Brahamaputra's headwaters is that the people up there don't surrender so easily, and you can never reckon the role that chance and plain luck will play in things. Also, because on that unhappy slog back to Mon, with all those ghosts on the road, Bilong said he knew someone with a vehicle that just might make it. And three hours later we were back where we'd given up in the Sumo, sliding and crawling up the same muddy hill, all packed into a miniature, antique Mahindra four-wheel-drive jeep. All of us, that is, except for Visevor, who was happy to wait back in Mon. We made it up the hill. It was mostly dumb luck, but we were back on the road to Longwa.

Our new driver was the owner of the Mahindra, a sharp little customer by the name of Phola. He'd driven a hard bargain for a Mahindra with no doors and a red plastic jerrycan in place of a gas tank; it bounced around on the floorboards with a hose coming out of it that disappeared under the hood. As the Mahindra struggled toward Mon, I sat half in and half out of the jeep, sometimes crouched on the running board. It was a bone-jarring, white-knuckle, head-bashing ride up through the canyons east of Mon, down again and across the Tegee River, over the top of the Veda Range, down into the Taphi Valley, then across the Taphi River and up through more steep canyons and switchbacks again. It went on like this for four hours, but we reached Longwa, up in the clouds at the top of Tainyai Mountain, in the late afternoon. A crowd of children came to greet us.

We made the final ascent on foot, up to a windswept peak and a massive structure roughly 40 metres long and 15 metres wide. The Angh of Longwa greeted us in his throne room, a dirt-floor chamber deep in the cavernous interior. But he was not what I had been expecting from the stories I'd been told. He was a young man and perfectly modern, in a blue windbreaker, a light blue shirt, grey slacks, and sandals. He sat on an old Assamese chair in front of a hearth fire in a chamber decorated with mithun skulls, elephant skulls, porcupine skulls, and several large gold-plated brass gongs. We made our introductions. His name was Donyei. He was 26 years old. It turned out that he was not the angh out of those stories, and no, he did not have 13 wives. Donyei spoke slowly, in Konyak, and Bilong did most of the translating.

First, there was Taiwang, Donyei said. It was probably the late sixteenth century. Taiwang was a Konyak prince from Pungchow, in what is now Arunachal Pradesh. After Taiwang came Lemwang, then Ngowang, Phowang, Ngowang II, Phowang II, Ngowang III, Phowang III, and Ngowang IV. Donyei's title, properly, was Phowang IV.

The angh with the 13 wives—the one who had beheaded those five disobedient noblemen—was Ngowang IV, Donyei's father. Donyei, the favoured son among his 30 children, had only recently assumed the throne, and he had only two wives. He laughed at something, and

Khrienuo, translating from the Nagamese, told me, he says he may have more wives some day, but for now, two is plenty.

I explained that I was a bit disoriented from the long and bumpy ride, and I couldn't even tell which way was north anymore. I asked about the border with Burma. How close was it? Bilong looked a bit confused. Burma, I said. You know, Myanmar. Bilong asked the question, then Donyei looked a bit confused, and there was some back and forth in Nagamese, with Khrienuo pitching in.

Then Bilong said, Oh, but you are in Burma.

It turned out that India's border with Burma, which was decided in 1963, wasn't actually surveyed around Tainyai Mountain until the early 1970s. Because the border ran along the height of land—across the top of Tainyai Mountain and straight through the village of Longwa—the main support post for the angh's palace was formally declared Border Post 158 of the Indo-Burmese frontier. The throne room happened to be on the Burmese side, the side where the sun rises. An Indian army lookout tower sat on a peak about two kilometres north, and the guards popped in to say hello from time to time. Every few weeks a Burmese military foot patrol would spend a couple of days in Longwa as it made its rounds, but none of these things were taken very seriously, Donyei explained. This was Naga territory, Konyak territory.

On the Burmese side, the closest Konyak villages were Langkho, Longkhai, and Khonmoi, and the nearest road in Burma was a seven-day walk away. The three closest villages on the Indian side were Pukha, Weting, and Nyanyu. They were all connected by footpaths, and at least 30 more villages still spoke the Longwa dialect and were nominally tributary to Longwa. Most of those villages were in Burma; others were in Arunachal Pradesh. But such borders mean nothing, Donyei said.

At this point in the conversation, Donyei's father, the renowned Ngowang IV, entered the throne room. He wore a brass amulet adorned with five figurine heads around his neck, and his hair was pulled into a tight topknot held in place with a monkey femur. We moved to the verandah of the longhouse to sit in the light of the dying sun.

Yes, I took those heads, he said, jingling his amulet. They were from Langkho village. I had to install a deputy there, he said. I think it was around 1970. There were territorial problems, he explained, and besides, they were taking heads from this village.

By this time, Khrienuo was letting Bilong do most of the translating, in that lovely birr-whing-bingbong cadence with which so many Naga languages seem to be blessed, and the conversation moved to the changes upon the world.

Donyei offered the opinion that there was nothing to worry about. The Konyak language would go on. His kingship would go on. The Longwa people would take from the world things of their own choosing, and life would unfold more or less as it always had.

His father, Ngowang IV, seemed less sure. Not worried, but not certain, either. The changes I have seen have been good, Ngowang IV said. In the old times, people lived in fear of one another. They hid from one another. But now there is peace. It is true that there are not so many animals, and the forest is changing, but in the old times the animals would come and kill our livestock, and they attacked our people. Now our people move throughout the mountains, and they have no fear of animals attacking them anymore.

I like living with lots of people around, he said. My fields can be cleared in one day. But if there are tigers, this is also a good thing. We kings of the Konyak people, our souls become tigers. Our spirits reside in the tigers, he explained. In one way, we do not like them. We do not want them to kill our mithuns and to kill our people. But if the tiger vanishes, our souls, too, will vanish.

I hadn't seen that coming. It was straight out of those old Malay stories about tigers being the manifestations of wandering human souls, or the ghosts of ancestors.

Ngowang IV then spoke slowly and carefully, and with unambiguous conviction. Bilong translated each sentence as he spoke.

We used to have many animals, he said. The boar, the deer, the royal stag—we had those. We had the chok, and the maya, and the villagers would go hunting and they would bring the meat to me. I was a deer hunter and a boar hunter and a leopard hunter. When I was a child, sometimes these animals could be killed with stones, and with branches from the trees. It is not that way now.

It troubles me, because if the tiger vanishes, our lives will vanish.

The lives of kings will vanish.

Epilogue
The Revenge of Kali

The formidable influence of the religious drive is based on far more ... than just the validation of morals. A great subterranean river of the mind, it gathers strengths from a broad spread of tributary emotions. Foremost among them is the survival instinct.

—Edward O. Wilson

In their last epic battle with the demons that were devouring the world, the gods of old sent streams of fire down from the Himalayas. The goddess Durga emerged from that fire, and Kali sprang from her brow and triumphed over the demons. Down through the centuries, around a stone image of Kali at an out-of-the-way jungle temple on the banks of a branch of the holy Ganges, a warren of shrines arose. The place came to be called Kalighat. The fetid and splendid city of Calcutta rose up around it, and Calcutta grew to become British India's imperial capital. Then the great river changed its course, and Kalighat was left to fester on the banks of the scum-covered Tollynalla Canal. There was a dead dog floating in it the first day I visited.

A young man held his infant son close to his chest and gingerly stepped through the mayhem of pilgrims in the alleys and corridors leading to the temple. The crowds moved slowly. Some of the pilgrims were in a state of panic. Others were perfectly serene. Everyone was draped in garlands of blood-red hibiscus. The frenzy rose and subsided among legions of bangle sellers and trinket wallahs, and it inched its way toward the epicentre of Kalighat, where Kali sits alone in a half-hidden chamber, with a garland of human heads around her neck.

The young man's name was Sibu Das. He was 23 years old, and he begged alms for a living. The child was his six-month-old son, Rahul. He and his wife, Kalpana, also had a three-year-old daughter, Sukla, and their home was a small patch of pavement outside Kalighat's main gate. At night they were allowed to sleep there undisturbed.

The precious little that Sibu Das could count among his family's blessings, such as the few rupees they gleaned in their daily routines, he credited to the patronage of Kali. A more secular reading would attribute the remnants of his family's meagre luck to the fact that they were among at least 500 beggars the temple priests at Kalighat served with meals every afternoon at four o'clock, and the mobile-battalion constables with the Calcutta Police are reluctant to roust beggars sleeping at the gates of temples.

When I met him, Sibu Das had been living with his family that way for two years. I'd paid a temple priest to be my interpreter during my time at Kalighat, and what I managed to gather from Sibu Das is that he had grown up on the other side of Tollygunge, the last station on the old Calcutta tramline, the youngest of five children. His father was a plumber who lost his job, and the family split up. Sibu never could find decent work, so he begged for alms at Kalighat. That was about all I could discern.

During the two years he and his family had been living on their patch of pavement outside Kalighat's main gate, roughly 25,000 goats had been ritually slaughtered at Kalighat, hundreds of thousands of people had been slain in wars in the Congo, Iraq, and Afghanistan, and elsewhere, and roughly 15 million of the world's people had died from hunger, malnutrition, and other such consequences of poverty. During those same two years, the world added to its human population about 150 million people. That's close to one-third of the number of all the human beings that were alive on earth when Kalighat was still a quiet, holy place in the jungle, in the late sixteenth century.

Still, on a good day, Sibu managed to beg rupees equal to $4 Canadian, and on those days he would be almost twice as well off as one-third of humanity, and so he would thank Kali. Those of us schooled in the disciplines of reason might sneer at such piety, but against the overwhelming evidence of modernity's colossal failure for people like Sibu Das and his family, all over the world, we would have no right at all. We have nothing to sneer about, not even in the face of the quackery at Kalighat, with its priests who make offers over the internet to perform pujas at Kali's feet for you, in absentia, at 674 rupees per puja. We have had our way, and the world has not been saved by Descartes or Darwin, Voltaire or Hobbes.

Throughout the twentieth century, the bright young secularists of the world sneered at the idiosyncratic superstitions of the poor, and the great coffin ship of earth sailed on, with the mysterious hidden hand of market forces at the helm. The hand sent down its enigmatic signals, and the technocrats in the engine room made appropriate adjustments to the calibrations of international trade rules. A virulent strain of Islam broke out across a vast stretch of the earth between the Pillars of Hercules and the Banda Sea, and Kali continued to draw her most loyal following among thieves, bandits, and the poor.

Among Kali's devotees, an organization known as Shiv Sena finds its most ardent recruits. In the 1980s, Shiv Sena was an irrelevant, obscure, off-the-deep-end relic from the Hindu–Muslim convulsions that followed India's partition, after independence. Twenty years later, Shiv Sena pretty much ran Bombay. Its thugs had come to control political life over a great swathe of the Indian subcontinent. The rise of Shiv Sena was one of those things nobody saw coming. It was like the changed world after September 11, 2001, the sudden disappearance of all those ancient and venerable food crops, and the talk about reviving the Tasmanian wolf from DNA extracted from a fetus in a pickle jar. So much is uncertain in the world, but one thing we can say with some certainty is that we are living in an age when we will at last discover the answer to the question that has haunted philosophers from time out of mind. It's the question about whether humanity is capable of determining its own destiny. We should know that by about 2030, they say. Certainly not much later.

For people like Sibu Das, all over the world, the Enlightenment's legacy can be read straight out of Jeremiah 8:20: *The summer is past, the harvest is ended, and we are not saved.* For the poor of the world, the harvest of modernity has come to mean the same "false promises of open sky and sea" Yevgeny Yevtushenko wrote about in his poems. For India's peasant farmers, it came in the form of hybrid seeds and the extinction of thousands of distinct varieties of indigenous rice. When their crippling indebtedness to seed companies was driving Indian farmers to suicide, *The Times of India* editorialized that what India's farmers really needed was a regimen of "anti-depressants ... taken under expert medical supervision." That's modernity for you.

To find some parallel with the conditions of recklessness and excess that prevail in the world, you have to scour such times as nineteenth-century Ireland in the days before the famine, or reach all the way back to the final days of the Sumerian, Roman, and Mayan empires. You have to look at those desperate moments in human history, the moments just before everything falls apart. You can also look ahead, and try to imagine where all this is leading. You can look at those United Nations Environment Programme *Global Outlook* scenarios again.

Wherever you look, there are choices. But the choices narrow every day, and they are becoming every bit as stark and momentous as the choices we faced in the darkest moments of the twentieth century, between fascism, totalitarianism, democracy, and socialism. Shaking our scythes at cannon, we answered those questions with a great deal of bloodshed. It would be nice to think we can avoid bloodshed in the decisions we have to make now, but bloodshed or not, we have to make those decisions. At least they're still available to us. That's the good news.

The big question facing humanity is not whether human beings will survive as a species. In the course of writing this book, I found no convincing evidence to support the popular contention that human behaviour is leading to our inevitable extinction. Even the worst scenarios of ecosystem loss, deforestation, species extinctions, global warming, and crop monoculture would not inevitably result in the extinction of humanity. We are an amazingly resilient species. The way American writer David Quammen puts it, we are a "weedy" species, like cockroaches and pigeons, and we are creating a planet fit only for weeds. But we are also Copernicus and Galileo. We are like cougars. We have figured out ways to colonize every ecological niche on earth. We will survive. So there's hope.

The choice is about *how* we will survive, and what kind of a planet we want for ourselves and for other living things. It is no longer good enough to talk about letting nature take its own course.

Humanity faces serious decisions, and serious decisions require that we believe in things, and believe deeply. Like humanity's love of living things and the way our brains are patterned for storytelling, our need to believe is ancient and abiding. Deep conviction, faith, ideology, religious devotion—all these things are fundamental, structural components of human nature. We have evolved that way; natural selection has favoured this capacity in us. We can sneer at the market fundamentalists of the world, despise the fascists terrorizing the Islamic part of it, and laugh at the pilgrims in their harmless devotions at Kalighat, but the one thing that has always moved people in times of great peril—the one thing that has allowed broad masses of people to determine their destiny—is the human capacity to believe in things.

In the absence of belief, we won't staunch the losses that extinction is tallying against the living, breathing world. We won't maintain the ancient legacies of language and literature and song in the world without a belief system that has some ancient, necessary, bloody thing at its centre, something every bit as grisly and stunning as that goddess at the fulcrum of the ecstasies at Kalighat.

E.O. Wilson, the "father of biodiversity" and one of the greatest minds of the twentieth century, has an idea about how the struggle for belief will unfold in the coming years. "The choice between transcendentalism and empiricism will be the coming century's version of the struggle for men's souls," he says. But the strongest beliefs derive from self-evident truths of the kind that reconcile the traditions of transcendentalism and empiricism. And there is a truth about extinctions, a thing that we all know to be true, somehow, to the very core of our being. In the way the poet William Blake put it, "Everything that lives is holy." You can speak those words with fire in the belly, or you can express them in a perfectly rational and secular manner. In United Nations Resolution A-RES-37-7, the World Charter for Nature, adopted October 28, 1982, the words appear this way: *Every form of life is unique, warranting respect regardless of its worth to man, and, to accord other organisms such recognition, man must be guided by a moral code of action.*

There have always been ways of saying this, in every language and culture on earth. You certainly don't need to rely solely on the lexicon of environmentalism to justify that ancient desire that persists in all of us to be in a living, breathing world, rich in the diversity and abundance of life. It is our right, and we should claim it, and humanity actually *is* capable of determining its own destiny. We certainly should not test that proposition by waiting to see if the ship sinks.

We should take the helm. We should reclaim the legacy of the Enlightenment for everyone, everywhere. We should reclaim the rights and entitlements of citizenship that have been stripped away from so much of the world. We can expand the scope of democracy, everywhere, and in ways that will allow us to confront the forces behind monoculture, ecological collapse, and all those other things that always seem to lead to a field where people make the sign of the cross when they pass. The work will require great sacrifice, discipline, and violence. We cannot shy away from the moral duty of that work because of a fear of those things. It will be hard work. It is certainly not the sort of thing you would want to leave to environmentalists.

Humanity has already had some brief moments at the helm, and they were exhilarating. In 1948, the founding member states of the United Nations drafted the Universal Declaration of Human Rights. That is something worth building upon, especially in a world faced with increasingly difficult questions about the protection of cultural diversity. When a great hole in the earth's paper-thin ozone layer was being ripped open by chlorofluorocarbons and halons, a handful of the world's nations signed on to a series of covenants to phase out those lethal substances, culminating in the Montreal protocol in 1997. By 2002, 183 nations had signed on. The United Nations' Kyoto protocol on climate change was an embarrassing baby step in the direction of addressing greenhouse-gas emissions. But people will only put up with so much. That's another conclusion I'd reached by the time I'd reached Kalighat. There's always a tipping point. It's just hard to predict when or where that happens, or upon whose back the last straw breaks.

The last thing I learned is a way to answer that question, "What, then, do we do?" It is this: You do what you can.

If it's some great insight you're after, all I can say is that the great insights lie only in the rich variety of humanity's stories, the specific and the particular stories, and the great multiplicity and diversity of our ideas. Our best hopes lie in strengthening the conditions that allow the flourishing of a diversity of living things, a diversity of ideas, and a diversity of choices.

Extinction is the thing that destroys those very conditions, so you join the epic battle with the demons that are devouring the world, and you do what you can. It's all anyone can expect of you. You do everything you can.

Notes and Sources

The epigraph, citing a passage from a letter George Owell wrote to Cyril Connolly in 1939, comes from Sonia Orwell and Ian Angus, eds., *The Collected Essays, Journalism and Letters of George Orwell, Volume I: An Age Like This, 1920–1940* (New York: Penguin Books, 1970).

Prologue: The Valley of the Black Pig

The *Guardian* article, "The way we'll live in 2032," from May 23, 2002, refers to United Nations Environment Programme, *GEO: Global Environmental Outlook 3: Past, Present and Future Perspectives* (Nairobi, Kenya: Division of Early Warning and Assessment, United Nations Environment Programme, 2002).

Yeats's poem "The Valley of the Black Pig" may be found in A. Norman Jeffares, ed., *Yeats: Selected Poetry* (London, UK: Macmillan, 1962).

For a broad overview of the patterns of species extinction through the earth"s history, see Niles Eldredge, *The Miner's Canary: Unraveling the Mysteries of Extinction* (New York: Prentice Hall, 1994). On the concept of the "sixth extinction," see F.S. Chapin, et al., "Consequences of Changing Biodiversity," *Nature,* May 11, 2000; Niles Eldredge, *Life in the Balance: Humanity and the Biodiversity Crisis* (Princeton, NJ: Princeton University Press, 2000); Niles Eldredge, "The Sixth Extinction," ActionBioscience.org., the American Institute for Biological Sciences online magazine, June 2001, www.actionbioscience.org/newfrontiers/eldredge2.html. On the "last critical reboubts" of biological diversity, see E.O. Wilson, *The Future of Life* (New York: Knopf, 2002).

The best online resource on the subject of threatened, endangered animals and plants and their status is the International Union for the Conservation of Nature (the World Conservation Union), at www.iucn.org, as well as the IUCN's "red list" of threatened species, at www.redlist.org.

Estimates that as many as a million species may be headed for extinction during the twenty-first century come from an analysis of the potential impacts of climate change in Chris D. Thomas, et al., "Extinction Risk from Climate Change," *Nature* 427, January 2004. For extinctions in food-plant varieties and domesticated animals, see Food and Agriculture Organization, *World Watch List for Domestic Animal Diversity*, 3rd ed. (Rome: FAO, 2000); Cary Fowler and Pat Mooney, *Shattering: Food, Politics and the Loss of Genetic Diversity* (Tucson: University of Arizona Press, 1990). For the Food and Agriculture Organization's online data about genetic erosion in food-plant varieties, see www.fao.org/ag/agp/agps/pgrfa/gpaeng.htm.

On the last known po`ouli: Jaymes Song, "Extinction near with native bird's death," *Honolulu Star-Bulletin,* December 1, 2004.

Two fine books about the state of the world's languages are Mark Abley, *Spoken Here: Travels among Threatened Languages* (Toronto: Random House, 2003), and Andrew Dalby, *Language in Danger: How Language Loss Threatens Our Future* (Toronto: Penguin, 2003). A rare acknowledgment of the relationship between the values at stake in both "natural" biological diversity and in artifacts of human culture, such as language and food varieties, is Beth Ann Fennelly, "Fruits We'll Never Taste," *Michigan Quarterly Review,* Fall 2000. A good overview of the statistical relationships among language, biological diversity, and food-crop variety is Tove Skutnabb-Kangas, et al., "Sharing a World of Difference: The Earth's Linguistic, Cultural and Biological Diversity" (Paris: UNESCO/World Wide Fund for Nature/Terralingua, 2003).

The idea that "nature" no longer exists as a place beyond the reach of humanity's impact on the planet was brilliantly developed by Bill McKibben in *The End of Nature* (New York: Random House, 1989). On "nature" as an objective reality and as a "constructed" reality, and on the controversy over "wilderness" preservation as a sensible response to ecological collapse, see William Cronon, ed., *Uncommon Ground: Rethinking the Human Place in Nature* (New York: W.W. Norton, 1996); Michael Soule and Gary Lease, eds., *Reinventing Nature? Responses to Postmodern Deconstructionism* (Washington, DC: Island Press, 1995).

For the proposition that environmentalism has outlived its usefulness, see Michael Shellenberger and Ted Nordhaus, "The Death of Environmentalism: Global Warming Politics in a Post-environmental World," January 13, 2005, www.thebreakthrough.org; Adam Werbach, "Environmentalism Is Dead: What

Next?" *In These Times,* June 21, 2005. The Nordhaus–Schellenberger paper prompted a vigorous debate in environmentalist circles. The debate was still unfolding at the online magazine *Grist* (www.grist.org) at the time this book was published.

On the desire for narrative patterned in the human brain, as Doris Lessing put it, see Robert Fulford, *The Triumph of Narrative: Storytelling in the Age of Mass Culture* (Toronto: House of Anansi Press, 1999). On the difficulty in "fortune-telling," especially in demographic changes in Western Europe, see Robert S. Leiken, "Europe's Angry Muslims," *Foreign Affairs,* July/August 2005. On exponential advances in technology and computing power: Joel Garreau, *Radical Evolution: The Promise and Peril of Enhancing Our Minds, Our Bodies, and What It Means to Be Human* (New York: Doubleday, 2004). On humanity's limited understanding of the sum of all living things: Colin Tudge, *The Variety of Life: A Survey and a Celebration of All the Creatures That Have Ever Lived* (Oxford, UK: Oxford University Press, 2000).

On humanity's expanding ecological footprint on earth, from the amount of methane we are pumping into the atmosphere to the percentage of the earth's primary productivity we take all to ourselves, see Bill McKibben, "A Special Moment in History," *Atlantic Monthly,* May 1998. See also Richard Manning, *Against the Grain: How Agriculture Has Hijacked Civilization* (New York: North Point Press/Farrar, Straus & Giroux, 2004); Richard Manning, "The Oil We Eat: Following the Food Chain Back to Iraq," *Harper's,* February 2004.

For the history of the countryside around Coolreagh, I am indebted to my uncle Tony and aunt Angela and my cousins Christine and Douglas, and also to Gerard Madden, *A History of Tuamgraney and Scarriff since Earliest Times* (Tuamgraney, Ireland: East Clare Heritage, 2000). For general Irish history, and the history of the "potato famine" of the mid-nineteenth century, see Edmund Curtis, *A History of Ireland* (London, UK: Methuen, 1936); Donald MacKay, *Flight from Famine: The Coming of the Irish to Canada* (Toronto: McClelland & Stewart, 1992).

Night of the Living Dead

Because the living world is undergoing such rapid change, the "slow" technology of a book is not too helpful in keeping abreast of the status of various endangered species. Change happens too quickly for anything but a constantly updated website. Most of that change is for the worse, but the trajectory of each species' condition is tracked haphazardly, depending on the resources and data available to various species specialists groups working with the International Union for the Conservation of Nature. See the IUCN's website (www.iucn.org), which contains links to the IUCN's Species Survival Commission and the IUCN's "red list" of species threatened with extinction. See also the eminently respectable Wildlife Conservation Centre (www.wcs.org).

On the Singapore Zoo, the Night Safari, and the Jurong Bird Park, see Singapore Zoological Gardens, "The World's First Night Safari" (n.d.); Abraham Verghese, "Singapore's Jurong Bird Park: A Study Model," *Resonance,* July 2001. For the "merlion" myth in the context of constructed identity and willed amnesia in Singapore, see William S.W. Lim, *Architecture, Art, and Identity in Singapore: Is There Life after Tabula Rasa?* (Singapore: Select Books, 2004). On zoo history in general: R.J. Hoage and William A. Deiss, eds., *New Worlds, New Animals: From Menagerie to Zoological Park in the Nineteenth Century* (Baltimore: Johns Hopkins University Press, 1996); particularly John C. Edwards, "The Value of Old Photographs of Zoological Collections."

On reckoning extinction rates: David Jablonski, "Extinction: Past and Present," *Nature* 427, February 12, 2004; Stuart Pimm, "Seeds of Our Own Destruction," *New Scientist* 146, April 8, 1995. On the estimated numbers of "living dead" species by 2050, see Thomas J. Foose, "Riders of the Last Ark: The Role of Captive Breeding in Conservation Strategies," in Les Kaufman and Kenneth Mallory, eds., *The Last Extinction* (Cambridge, MA: MIT Press, 1986); "Singapore slams media watchdog for low ranking in press freedom," *Malaysia Star,* November 17, 2004.

For the emergence of chimeras, artificial intelligence, cyborgs, and so on, see Freeman Dyson, "The Darwinian Interlude," *MIT Technology Review,* March 2005; H.G. Wells, *The Island of Doctor Moreau* (New York: Penguin Putnam, 1988); Joel Garreau, *Radical Evolution: The Promise and Peril of Enhancing Our Minds, Our Bodies, and What It Means to Be Human* (New York: Doubleday, 2004); Linda MacDonald Glenn, "Legal and Ethical Issues in Transgenics and the Creation of Chimeras," Walter C. Randall Biomedical Ethics Lecture, San Diego, CA, 2003; Hannah Kamenetsky, "Chimera

Conflict: Human–Mouse Hybrid Plan Faces Patent Challenge from Activist Rifkin," *The Scientist,* December 10, 2002.

For daunting questions about the diversity of life on earth and how science makes sense of it, see Colin Tudge, *The Variety of Life: A Survey and a Celebration of All the Creatures That Have Ever Lived* (Oxford, UK: Oxford University Press, 2000); E.O. Wilson, *The Future of Life* (New York: Knopf, 2002).

An amusing reference for newly discovered and rediscovered animals is Matthew A. Bille, *Rumors of Existence: Newly Discovered, Supposedly Extinct, and Unconfirmed Inhabitants of the Animal Kingdom* (Surrey, BC: Hancock House, 1995).

See also William Newmark, "A Land-Bridge Island Perspective on Mammalian Extinctions in Western North American Parks," *Nature* 325, January 29, 1987; Peter Boomgaard, *Frontiers of Fear: Tigers and People in the Malay World, 1600–1950* (New Haven, CT: Yale University Press, 2001); Barry W. Brook, Navjot S. Sodhi, and Peter K.L. Ng, "Catastrophic Extinctions Follow Deforestation in Singapore," *Nature* 424, July 24, 2003; G. Cowlishaw, "Predicting the Pattern of Decline of African Primate Diversity: An Extinction Debt from Historical Deforestation," *Conservation Biology* 13, 1999; Vicki Croke, *The Modern Ark: The Story of Zoos, Past, Present and Future* (New York: Scribner, 1997); David Hancocks, *A Different Nature: The Paradoxical World of Zoos and Their Uncertain Future* (Berkeley/Los Angeles: University of California Press, 2001); David Malamud, *Reading Zoos: Representations of Animals in Captivity* (New York: New York University Press, 1998); Ian Sample, "Frozen ark to save rare species: DNA of ancient creatures to be kept for posterity," *The Guardian,* July 27, 2004; Mark Schulman, "Revive the extinct Tasmanian tiger—through cloning?" *Christian Science Monitor,* July 11, 2002.

Waiting for the Macaws

For the account of the last great auk at St. Kilda, as well as many accounts of the rare and now extinct birds of the earth, see James C. Greenway, Jr., *Extinct and Vanishing Birds of the World* (New York: Dover Publications, 1962).

For the current status of birds throughout the world, see www.iucn.org and www.redlist.org. Birdlife International maintains an extensive and detailed database of the world's birds at www.birdlife.org.

For an overview of the world's parrots, including the macaws: John Sparks and Tony Soper, *Parrots: A Natural History* (New York: Facts On File, 1990). For an overview of the threats to the world's macaws, see the World Wildlife Fund's research and information section for macaws at www.wwf.org.uk/core/wildlife/fs_0000000024.asp. The Convention on the International Trade in Endangered Species (CITES) database consulted is *UNEP–WCMC Species Database: CITES-Listed Species,* August 4, 2005, www.cites.org/eng/resources/species.html.

Helmut Sick on the "blue" macaws: "About the Blue Macaws, Especially the Lear's Macaw," *Conservation of New World Parrots,* Proceedings of the International Council for Bird Preservation (ICBP) Parrot Working Group Meeting in St. Lucia, 1980 (the ICBP is now Birdlife International, Wellbrook Court, Girton Road, Cambridge CB3 0NA, UK). And on Helmut Sick: François Vuilleumier, "In Memoriam: Helmut Sick, 1910–1991," *The Auk* 115, no. 2, 1998.

Tony Pittman is an avid macaw fancier, a writer, and a tireless researcher and crusader who is probably as knowledgeable as anyone alive about macaws, especially "blue" macaws. He maintains a regularly updated website at www.bluemacaws.org.

British conservationist Tony Juniper took time off from his duties as the executive director of the environmental organization Friends of the Earth to tell the story of the Spix's macaw in a profoundly moving work that significantly informed the sections of this chapter about the recovery program for the species: *Spix's Macaw: The Race to Save the World's Rarest Bird* (London: Fourth Estate, 2002). On Qatar's Sheikh Saoud bin Mohammed bin Ali al Thani and his interest in Spix's macaws: Al Wabra Wildlife Preservation, press release: "Report on the current situation of the Spix Macaw *(Cyanopsitta spixii)* population in the care of Al Wabra Wildlife Preservation (AWWP), State of Qatar," July 12, 2004. The Loro Parque breeding program for Spix's macaws has a website at www.loroparque-fundacion.org.

One study of the scarlet macaw reintroduction program at the Curú National Wildlife Refuge is Donald Brightsmith, Jennifer Hilburn, et al., "The Use of Hand-Raised Psittacines for Reintroduction: A Case Study of Scarlet Macaws *(Ara macao)* in Peru and Costa Rica," *Biological Conservation* 121, 2005. On the Carolina parakeet, the ivory-billed woodpecker, the passenger pigeon, and the great auk, Americans' love of birds, and the sorry behaviour of Americans who did not love them: Christopher Cokinos, *Hope Is the Thing with Feathers: A Personal Chronicle of Vanished Birds* (New York: Warner Books, 2000).

For the "surprised enchantments" associated with birds and the great pastime of birdwatching, see Joseph Kastner, *A World of Watchers* (New York: Knopf, 1986). On the last days of Vancouver's crested mynahs, see Larry Pynn, "Last of city's non-native mynahs die," *The Vancouver Sun,* March 1, 2003. And on bird extinctions in North America in general, and the struggles to prevent extinction, as well as early conservation efforts for other species: Peter Matthiessen, *Wildlife in America* (New York: Viking Press, 1959).

For the dodo and its demise, and the complexities involved in the extinction of island species and isolated species, see David Quammen, *The Song of the Dodo: Island Biogeography in an Age of Extinction* (New York: Touchstone Books, 1996). On the last days of the great auk, especially at Funk Island: Farley Mowat, *Sea of Slaughter* (Toronto: McClelland & Stewart, 1984). On market hunting, the role of the wild game industry, and conservation efforts in Canada, see George Colpitts, *Game in the Garden: The Human History of Wildlife in Western Canada to 1940* (Vancouver: University of British Columbia Press, 2002). And on the shatoosh shawl: Yvonne Roberts, "Butchered by the score: The bloody and deadly trade that has brought a much-prized antelope to the brink of extinction," *The Observer,* March 21, 2004.

Costa Rica's distinct political traditions and its "sustainability first" focus is discussed in Jeff Langholz, et al., "Incentives for Biological Conservation: Costa Rica's Private Wildlife Refuge Program," *Conservation Biology* 14, December 2000; Eugene D. Miller, *A Holy Alliance? The Church and the Left in Costa Rica, 1932–1948* (Armonk, NY: M.E. Sharpe, 1996); A. Sanchez-Azofeifa, et al., "Integrity and Isolation of Costa Rica's National Parks and Biological Reserves: Examining the Dynamics of Land-Cover Change," *Biological Conservation* 109, 2003. On the rich avifauna of Costa Rica, see Gary F. Stiles and Alexander Skutch, *A Guide to the Birds of Costa Rica* (Ithaca, NY: Comstock Publishing/Cornell University, 1989).

For the rediscovery of the ivory-billed woodpecker, see Cornell University Ornithology Lab, "Long Feared Extinct, a Magnificent Bird Still Lives," www.birds.cornell.edu/ivory; John J. Fitzpatrick, et al., "Ivory-Billed Woodpecker *(Campephilus principalis)* Persists in Continental North America," *Science,* June 3, 2005. And on the rediscovery controversy and its apparent conclusion: James Gorman and Andrew C. Revkin, "Vindication for ivory billed woodpecker and its fans," *The New York Times,* August 2, 2005.

The Last Giants in the River of the Black Dragon

It may be that I haven't done justice to the Wild Salmon Center's ideas for saving "the last of the best" of the world's salmon rivers by setting them apart in special protected zones. I should say here that the Portland-based organization does tremendously generous work in the Russian Far East. The centre has become a vital source of support for Russia's best and most honest scientists and conservationists. It was the Wild Salmon Center that invited me to the Khabarovsk symposium where much of this chapter unfolds (I presented a paper there about salmon conservation in Canada, on behalf of the Sierra Club of Canada's B.C. chapter), and the centre co-sponsored the symposium with Ecodal (Khabarovsk) and the Russian government's Institute of Water and Ecological Problems. The Wild Salmon Center's website is at www.wildsalmoncenter.org.

The epigraph, from the poem "Ballad about False Beacons," is from Yevegeny Yevtushenko, *Stolen Apples* (New York, Doubleday & Co., 1972).

On illegal logging operations in the Russian Far East, see Steve Connor, "Fiddling while Siberia burns: 'Lungs of Europe' under threat from forest fires." *The Independent,* May 31, 2005.

I relied on a series of rather technical and academic papers for background research on the Amur River and its fisheries. Unfortunately, many of the documents I used were either translated in summary from the Russian or are otherwise unavailable. For scholarly information on Sakhalin taimen, for instance, the only substantial work is "Taimens and Lenoks of the Russian Far East," by S. Zolotukhin, A. Semenchenko, and V. Belyaev (2000), but it is in Russian. For background on the fisheries resources of the Amur River, see Yuri Darman and Eugene Simonov, *Amur/Heiliong River Basin Initiative* (Vladivostok: World Wildlife Fund Russia, Far Eastern Branch, 2002); Yu V. Slynko, *Fish Biodiversity of the Amur River* (Borok, Russia: Institute for Biology of Inland Waters RAS, n.d.). The ethnography and oral traditions of the Amur River peoples are discussed in Kira Van Deusen, *The Flying Tiger: Women Shamans and Storytellers of the Amur* (Montreal: McGill-Queen's University Press, 2001).

On Russia's tragic descent into gangsterism, see David Satter, *Age of Delirium: The Decline and Fall of the Soviet Union* (New Haven, CT: Yale University Press, 1996). For the criminal control of Russian industries, gangsterism in Vladivostok, and the starvation of Russian sailors: David Satter, *Darkness at Dawn: The Rise of the Russian Criminal State* (New Haven, CT: Yale University Press, 2003). On the "shock therapy" Boris Yeltsin and his successors visited upon the Russian people, see Jean-Marie Chauvier, "Russia: nostalgic for the Soviet era," *Le Monde Diplomatique,* March 2004; Michael Specter, "Is Russia Dying?" *The New Yorker,* October 11, 2004.

The Khabarovsk public opinion survey undertaken by the Khabarovsk Wildlife Federation is reported in A.S. Sheinhouse and G.I. Sukhomirov, "Evaluation of the Population of the Problems Facing Specially Protected Territories in the Mountain Forests of Sikhote Alin" (Khabarovsk: Khabarovsk Wildlife Federation, 2002). The B.C. survey is reported in Mel Kotyk and Adam Di Paula, *Benchmark Assessment of Public Awareness, Knowledge, Attitudes and Behaviour* (Vancouver, BC: Habitat Conservation and Stewardship Program, Fisheries and Oceans Canada, April 2000).

For Vladimir Putin's attitude toward ecological protection, see Rory Cox, "Putin sets back ecological clock," *Pacific Environments Newletter* 2, Summer 2000; Mark Hertsgaard, "Russia is an eco-disaster, and it just got worse," *The Washington Post,* July 9, 2000.

For the collapse of order in the Russian Far East, including the mass poaching of rivers, illegal logging, the violation of laws protecting "nature reserves," and the dismanting of those reserves, I relied heavily on the presentations to the Khabarovsk symposium. Most of these presentations are gathered in the proceedings: *North Pacific Salmon Protected Areas Workshop May 6–8, 2003, Khabarovsk, Russia* (Portland, OR: Wild Salmon Center). Those proceedings include Sergei Makeev's description of the Chernomyrdon bear hunt, in "Krilyon Suffering." The problems associated with establishing, maintaining, and protecting "nature reserves" and other such parks and protected areas for the conservation of salmon and other resources were discussed by A.L. Antonov (Institute of Water and Environmental Problems, Far Eastern Branch of Russian Academy of Science, Khabarovsk) in "Analysis of the Protected Areas Situated in the Khabarovsk Part of the Amur River Watershed and Ways to Save Salmon"; Irina Borisovna Bogdan (Far Eastern Ecodal), in "Laws of Creation and Functioning of Protected Areas for

Conservation of Salmon and Its Habitat"; S.N. Safronov, et al. (Laboratory of Hydroecology, Sakhalin State University), in "Nature Protection Areas of Sakhalin and Conservation of Rare Salmonids"; S.F. Zolotukhin (TINRO-Centre, Khabarovsk), in "Protected Areas and Sakhalin Taimen in Khabarovsk Krai"; and S.F. Zolotukhin and A.V. Shishaev (TINRO-Centre, Khabarovsk), in "Sakhalin Taimen Populations in Rivers of the Western Coast of the Tartar Strait and Possibilities for Rational Use." Overall problems in fisheries management along the Amur River were discussed at the symposium by A. Darman, E.A. Simonov, and P.O. Sharov (WWF Russia, Far Eastern Branch, Vladivostok), in "Integrated River Basin Management in Amur Watershed."

For kaluga sturgeon poaching on the Amur, see Anatoly Medetsky, "Black caviar feeds thousands on Amur," *Vladivostok News,* September 15, 2000. For problems associated with the salmon runs of the Sikhote Alin mountains, see the symposium paper by I.Z. Parpura (Scientific Research Station, TINRO-Centre, Ternei), "Salmon of the Eastern Slopes of the Sikhote Alin: Problems of Protection and Some Ways to Solve Them."

On specific hardships facing the aboriginal peoples of the Russian Far East, see the symposium papers, T.S. Khetani (Indigenous Small Nations of the North of Magadan Oblast "Ai Bini" Union), "Salmon and Indigenous Communities in Magadan"; G.I. Sykhomirov and V.A. Belyaev (Institute of Economic Research, Department of TINRO-Centre, Khabarovsk), "Role of Native Peoples of the North in the Creation of Salmon Protected Areas and Opportunities for Using Territories of Traditional Land Use for Salmon Preservation."

On the collapse of order in Kamchatka, especially with respect to illegal bear hunting, see Kim Murphy, "The vanishing bears: Economics vs. nature in the Russian wilds," *The Los Angeles Times,* November 27, 2003. On the Aral Sea disaster: Tom Bissell, *Chasing the Sea: Lost among the Ghosts of Empire in Central Asia* (New York: Pantheon Books, 2003); Rob Ferguson, *The Devil and the Disappearing Sea* (Vancouver: Raincoast Books, 2003). The collapse of saiga populations is described in Laura Williams, "Saga of the Saiga: The Hump-Nosed Antelope of the Russian and Central Asian Steppe May Soon Disappear," *National Wildlife* (National Wildlife Federation), April–May 2004.

Trends in global fisheries catches are outlined in R.J.R. Grainger and S.M. Garcia, *Chronicles of Marine Fishery Landings (1950–1994): Trend Analysis and Fisheries Potential,* FAO Fisheries Technical Paper No. 359

(Rome: Food and Agriculture Organization, 1996); Ransom A. Myers and Boris Worm, "Rapid Worldwide Depletion of Predatory Fish Communities," *Nature* 423, May 15, 2003; Daniel Pauly, et al., "The Future for Fisheries," *Science* 302, no. 21, November 2003.

Nineteenth-century life along the Amur River is described in Perry McDonough Collins, *Siberian Journey down the Amur to the Pacific, 1856–1857,* edited by Charles Vevrier (Milwaukee: University of Wisconsin Press, 1962). For an overview of the aboriginal peoples of the North Pacific, see William W. Fitzhugh and Aron Crowell, eds., *Crossroads of the Continents: Cultures of Siberia and Alaska* (Washington, DC: Smithsonian Institution, 1988).

The Ghost of the Woods

The epigraph, quoting William Hornaday, can be found in Peter Matthiessen, *Wildlife in America* (New York: Viking Press, 1959), 180.

Known causes of extinctions since 1600 are outlined in B. Groombridge, ed., *Global Biodiversity: Status of the Earth's Living Resources* (London: Chapman and Hall/World Conservation Monitoring Centre, 1992). For an investigation of the place that "man-eating" predators occupy in the physical world and in the human consciousness, see David Quammen, *Monster of God: The Man-Eating Predator in the Jungles of History and the Mind* (London: Norton, 2003).

A general overview of the causes and consequences of the dramatic rise in cougar attacks in North America is Terry Glavin, "Cougar Attack: Surviving an Ambush in Big Cat Country," *Canadian Geographic,* May 2004. For the story of the cougar in North America: Chris Bolgiano, *Mountain Lion: An Unnatural History of Pumas and People* (Mechanicsburg, PA: Stackpole Books, 1995); Harold P. Danz, *Cougar* (Athens, OH: Swallow Press/Ohio University Press, 1999); R.D. Lawrence, *The Ghost Walker* (New York: Holt, Rinehart and Winston, 1983); Gerry Parker, *The Eastern Panther: Mystery Cat of the Appalachians* (Halifax, NS: Nimbus Publishing, 1998); J. Bob Tinsley, *The Puma: Legendary Lion of the Americas* (El Paso: Texas Western Press/University of Texas, 1987); Bruce S. Wright, *The Eastern Panther: A Question of Survival* (Toronto: Clarke, Irwin, 1972); Bruce S. Wright, *The Ghost of North America* (New York: Vantage Press, 1959).

The late-twentieth-century rise in cougar attacks in North America is best documented in Paul Beier, "Cougar Attacks on Humans in the United States and Canada," *Wildlife Society Bulletin* 19, 1991; and Jo Deurbrouck and Dean Miller, *Cat Attacks: True Stories and Hard Lessons from Cougar Country* (Seattle, WA: Sasquatch Books, 2001). Recent trends in cougars in the eastern parts of North America are described in Roderick E. Cumberland and Jeffrey A. Dempsey, "Recent Confirmation of a Cougar, *Felis concolor,* in New Brunswick," *The Canadian Field Naturalist* 108, 1994; D.S. Maehr, et al., "Eastern cougar recovery is linked to the Florida panther: Cardoza and Langlois revisited," *Wildlife Society Bulletin* 31, no. 3, 2003. For the evolutionary history of cougars, see M. Culver, et al., "Genomic Ancestry of the American Puma *(Puma concolor),*" *The Journal of Heredity* 91, no. 3, 2000.

For recent advances in our understanding of the peopling of the New World, see J.M. Adovasio, *The First Americans: In Pursuit of Archeology's Greatest Mystery* (New York: Random House, 2002); Brian M. Fagan, *The Great Journey: The Peopling of Ancient America* (London: Thames and Hudson, 1987); Tom Koppel, *Lost World: Rewriting Prehistory—How New Science Is Tracing America's Ice Age* (Markham, ON: Simon and Schuster, 2003). On the pre-European population of the New World and the pre-European cultures' shaping of ecoysystems, see William R. Iseminger, "Mighty Cahokia," *Archeology* 49, May/June 1996; Charles C. Mann, "1491," *The Atlantic Monthly,* March 2002.

For John Muir, the Hohokam, and California's early palm oases, see Gary Paul Nabhan, "Cultural Parallax in Viewing North American Habitats," in Michael Soule and Gary Lease, eds., *Reinventing Nature? Responses to Postmodern Deconstructionism* (Washington, DC: Island Press, 1995). On the "mountain beaver," see Terry Glavin, "Haplodons and others pushed into oblivion," *The Georgia Straight,* May 24–31, 2001; L.W. Gyug, "Status, Distribution, and Biology of the Mountain Beaver, *Aplodontia rufa,* in Canada," *Canadian Field-Naturalist* 114, July–September 2000.

For an overview of human-caused extinctions, see Jean Christophe Balouet, *Extinct Species of the World: 40,000 Years of Conflict* (New York: Barron's, 1990). On the complexity of human-caused extinctions over the past 500 years, and North American attitudes toward wildlife and conservation: George Colpitts, *Game in the Garden: The Human History of Wildlife in Western Canada to*

1940 (Vancouver: University of British Columbia Press, 2002); Peter Matthiessen, *Wildlife in America* (New York: Viking Press, 1959); Farley Mowat, *Sea of Slaughter* (Toronto: McClelland & Stewart, 1984); Roderick Nash, *Wilderness and the American Mind* (New Haven, CT: Yale University Press, 1967); David Quammen, *The Song of the Dodo: Island Biogeography in an Age of Extinction* (New York: Touchstone Books, 1996).

On Early Holocene extinctions in North America, see John Elroy, "A Multispecies Overkill Simulation of the End-Pleistocene Megafaunal Mass Extinction," *Science* 292, no. 8, June 2001; Tim Flannery, *The Eternal Frontier: An Ecological History of North America and Its Peoples* (New York: Atlantic Monthly Press, 2001). For an opposing view to Elroy's, see Barry W. Brook and David M.J.S. Bowman, "Explaining the Pleistocene Megafaunal Extinctions: Models, Chronologies and Assumptions," *Proceedings of the National Academy of Sciences,* November 12, 2002. For further evidence of the human hand in Early Holocene extinctions, see Jean Christophe Balouet and Storrs L. Olson, "Fossil Birds from Late Quaternary Deposits in New Caledonia," *Smithsonian Contributions to Zoology,* January 10, 1989; Betsy Mason, "Big Beast Extinctions Blamed on Prehistoric Fire Starters," *New Scientist,* August 10, 2003; David A. Burney, Guy S. Robinson, and Lida Pigott Burney, "Sporormiella and the Late Holocene Extinctions in Madagascar," *Proceedings of the National Academy of Sciences,* September 16, 2003; Gary Haynes, "The Catastrophic Extinction of North American Mammoths and Mastodons," *World Archeology* 33, no. 3, 2002; Charles E. Kay, "Aboriginal Overkill and Native Burning: Implications for Modern Ecosystem Management," *Western Journal of Applied Forestry* 10, no. 4, 1995; O. Ludovic, et al., "Ancient DNA Analysis Reveals Woolly Rhino Evolutionary Relationships," *Molecular Phylogenetics and Evolution* 28, 2003; G.H. Miller, et al., "Pleistocene Extinction of *Genyornis newtoni:* Human Impact on Australian Megafauna," *Science* 283, 1999; David W. Steadman and Paul S. Martin, "The Late Quaternary Extinction and Future Resurrection of Birds on Pacific Islands," *Earth Science Reviews* 61, 2003; S.L. Vartanya, V.E. Garutt, and A.V. Sher, "Holocene Dwarf Mammoths from Wrangel Island in the Siberian Arctic," *Nature* 362, 1993.

On mule deer, see Valerius Geist, *Deer of the World: Their Evolution, Behavior, and Ecology* (Mechanicsburg, PA: Stackpole Books, 1998).

Drifting into the Maelstrom

The poem about the Maelstrom, by Petter Dass, is from Trygve-Olaf Lindvig, *The Maelstrom of Moskenes in Poetry and Stories* (Lillehammer, Norway: Natura Forlag, n.d.). Other sources for the Maelstrom: B. Gjevik, H. Moe, and A. Ommundsen, "Strong Topographic Enhancement of Tidal Currents: Tales of the Maelstrom," *Proceedings,* Norwegian Academy of Science and Letters, 1997; W. Ley, "Scylla Was a Squid," *Natural History Magazine,* June 1941.

On the status of minke whales, see Cetacean Specialist Group 1996, *"Balaenoptera acutorostrata,"* in *IUCN Red List of Threatened Species,* August 19, 2005, www.redlist.org.

A remarkable number of books have been written about the exploits of Captain Paul Watson and the Sea Shepherd Society, but for the greatest insight, see the books Watson has written or co-authored: Paul Watson, *Earthforce: An Earth Warrior's Guide to Strategy* (La Canada, CA: Chaco Press, 1993); Paul Watson, *Ocean Warrior: My Battle to End the Illegal Slaughter on the High Seas* (Toronto: Key Porter Books, 1994); Paul Watson and Warren Rogers, *Sea Shepherd: My Fight for Whales and Seals* (New York: W.W. Norton, 1982).

For a summary of the Makah whaling controversy, see Terry Glavin, "Beyond the grey whales," *The Globe and Mail,* October 9, 1988. On the history of industrial whaling, see Daniel Francis, *A History of World Whaling* (Toronto: Penguin, 1991).

A good general history of the Norse is Rolf Danielsen, et al., *Norway: A History from the Vikings to Our Own Times,* translated by Michael Drake (Oslo, Norway: Scandinavian University Press, 1995). For Norwegian and arctic whaling, from antiquity through the riots of the early twentieth century, see Igor Krupnik, *Arctic Adaptations: Native Whalers and Reindeer Herders of Northern Eurasia,* translated by Marcia Levenson (Lebanon, NH: University Press of New England, 1993); E.O. Oeein, "Coastal Hunting off Norway in the Old Days," *Norsk Veterinaertidsskr* 109, 1997. For Norwegian whaling from the 1860s to the resumption of commercial whaling under the IWC's "exception" rule, and for descrimination against "non-aboriginal" small whalers, see R. Gambell, "Management of Whaling in Coastal Communities," Arne Kalland, "Marine Mammals in the Culture of Norwegian Coastal Communities," and E.O. Oeein, "Norwegian Whaling," all in *Whales, Seals,*

Fish and Man, proceedings of the International Symposium on the Biology of Marine Mammals in the North East Atlantic (Tromsø, Norway: 1994); Will Plutte, "The Whaling Imperative: Why Norway Whales," *Oceans,* March 1984.

Controversies surrounding Norwegian, Icelandic, and other whale hunts in the North Atlantic are discussed in Gudrun Petursdottir, ed., *Whaling in the North Atlantic: Economic and Political Perspectives* (Rejkjavik: Fisheries Research Institute/University of Iceland Press, 1997). On the cultural origins of the "Save the Whales" movement, see Robert Hunter, *The* Greenpeace *to Amchitka: An Environmental Odyssey* (Vancouver: Arsenal Pulp Press, 2004); Rex Weyler, *Greenpeace: How a Group of Ecologists, Journalists and Visionaries Changed the World* (Vancouver: Raincoast Books, 2004). For the environmentalist "totemization" of whales, A. Kalland, "Management by Totemization: Whale Symbolism and the Anti-Whaling Campaign," *Arctic* 46, no. 2, 1993.

On the "Western" tradition of care for non-human forms of life, see Rod Preece, *Animals and Nature: Cultural Myths, Cultural Realities* (Vancouver: University of British Columbia Press, 1999); Keith Thomas, *Man and the Natural World: Changing Attitudes in England, 1500–1800* (London: Penguin, 1983).

An objective assessment of the question of a "special" cetacean intelligence may be found in Margaret Klinowksa, "Brains, Behaviour and Intelligence in Cetaceans (Whales, Dolphins and Porpoises)," *11 Essays on Whales and Man* (Reine i Lofoten, Norway: High North Alliance, 1994). The argument in favour of a special cetacean intelligence, including Carl Sagan's commentary, John Lilly's research, whales' responses to live rock music, etc., is represented in Joan McIntyre, ed., *Mind in the Waters: A Book to Celebrate the Consciousness of Whales and Dolphins* (Toronto: McClelland & Stewart, 1974). For humane issues regarding contemporary whale hunts, see S.C. Kestin, "Review of Welfare Concerns Relating to Commercial and Special Permit (Scientific) Whaling," *The Veterinary Record,* March 10, 2001.

On such cultural eccentricities as Lofoteners' affection for eider ducks, see "Hard to Swallow," *Marine Hunters: Whaling and Sealing in the North Atlantic* (Reine i Lofoten, Norway: High North Alliance, 1997).

For controversies regarding "aboriginal" and "commercial" whaling in the International Whaling Commission regime, see Milton M.R. Freeman, "Is Money the Root of the Problem?" in Robert L. Friedheim, ed., *Towards a*

Sustainable Whaling Regime (Seattle: University of Washington Press, 2000); Arne Kalland, "Aboriginal Subsistence Whaling: A Concept in the Service of Imperialism," *11 Essays on Whales and Man* (Reine i Lofoten, Norway: High North Alliance, 1994); R.R. Reeves, "The Origins and Character of Aboriginal Subsistence Whaling: A Global Review," *Mammal Review* 32, 2002); H.N. Scheiber, "Historical Memory, Cultural Claims and Environmental Ethics in the Jurisprudence of Whaling Regulation," *Ocean and Coastal Management* 38, no. 1, 1998.

On the International Whaling Commission's forcing Inuit to "live out the fantasies of white people about Eskimos," see "Greenland Whalers Demand: The Right to Be Commercial," *The International Harpoon* (High North Alliance, July 25, 2001). For an overview of the International Whaling Commission, its disputes and controversies: Peter J. Stoett, *The International Politics of Whaling* (Vancouver: University of British Columbia Press, 1997). Convention on the Trade in Endangered Species (CITES) secretary general Willem Winjstekers's protests about the conduct of the IWC are reported in "IWC Chairs Mislead CITES," *The International Harpoon* (High North Alliance, July 25, 2001).

For the "Irish proposal" to resolve controversies of sustainable coastal whaling, see "Ireland Bids for Nobel Peace Prize," *The International Harpoon* (High North Alliance, October 20, 1997). Banks accuses Brundtland of being a "murderer" for supporting whaling: "On the Whalers," *The International Harpoon* (High North Alliance, July 23, 2001). IWC chairman Phil Hammond's resignation, and his explanation: "Small Time Whaling and Big Time Politics," *Living off the Sea: Minke Whaling in the Northeast Atlantic* (Reine i Lofoten, Norway: High North Alliance, 1994). Gro Harlem Brundtland's letter supporting Norwegian minke-whale hunting: *Marine Hunters: Whaling and Sealing in the North Atlantic* (Reine i Lofoten, Norway: High North Alliance, 1997).

On the difficulties "whale preservation" poses for international efforts to conserve marine species, see J.M. Hutton and B. Dickson, "Conservation out of Exploitation: A Silk Purse from a Sow's Ear?" in *Conservation of Exploited Species, Conservation Biology* 6 (special issue), 2002. For Lofoten folk beliefs, see Sherry Von Ohlsen-Karasik, "Giving Waters," *World & I,* 8, no. 12, August 1997. On the difficulties posed for African peoples in "nature-for-development" swaps, see John Vidal, "Forced from rich forests into squalour," *The Guardian Weekly,* October 2–8, 2003; John Vidal, "Ousted from Africa,"

The Guardian, August 21, 2003. Also see the various publications of the Forest Peoples Programme: www.forestpeoples.org.

Edgar Allan Poe's *Descent into the Maelstrom* comes from Trygve-Olaf Lindvig, *The Maelstrom of Moskenes in Poetry and Stories* (Lillehammer, Norway: Natura Forlag, n.d.).

An Apple Is a Kind of Rose

The epigraph for this chapter comes from Pat Mooney, a pioneering writer and researcher on the topic of vanishing food-crop diversity: Pat Roy Mooney, *The Parts of Life: Agricultural Biodiversity, Indigenous Knowledge, and the Role of the Third System* (Uppsala, Sweden: Dag Hammarskjold Centre, 1998), 14.

Throughout this chapter, for data on the decline in food-crop diversity, I relied heavily on *The Parts of Life* and others of Mooney's works, such as Cary Fowler and Pat Mooney, *Shattering: Food, Politics and the Loss of Genetic Diversity* (Tucson: University of Arizona Press, 1990); Pat Roy Mooney, *The Law of The Seed: Another Development and Plant Genetic Resources* (Uppsala, Sweden: Dag Hammarskjold Centre, 1983); and Pat Roy Mooney, et al., *The Laws of Life: Another Development and the New Biotechnologies* (Uppsala, Sweden: Dag Hammarskjold Centre, 1988).

On apples and potatoes, see Michael Pollan, *The Botany of Desire* (New York: Random House, 2001). Several overviews of the rapid decline in the diversity of food crop varieties are available at www.slowfoodfoundation.com.

On language loss, see Andrew Dalby, *Language in Danger: How Language Loss Threatens Our Future* (Toronto: Penguin, 2003). For global deforestation: Derrick Jensen and George Draffan, *Strangely Like War: The Global Assault on Forests* (White River, VT: Chelsea Green Publishing, 2003). For habitat loss and its relation to extinctions: International Union for the Conservation of Nature, "Red List of Threatened Plants" (Cambridge, UK: IUCN Publications Service Unit, 1997). On the loss of domesticated animals: Sustainable Development Department, Food and Agriculture Organization, "Biodiversity for Food and Agriculture: Farm Animal Genetic Resources" February 1998, www.fao.org/sd/EPdirect/EPre0042.htm.

The "Sitka Biome," its plant resources, and language diversity are described in Ecotrust, Pacific GIS, and Conservation International, *The Rain Forests of Home: An Atlas of People and Place* (Vancouver: Ecotrust Canada, 1995); Erna

Gunther, *Ethnobotany of Western Washington* (Seattle: University of Washington Press, 1945); Hilary Stewart, *Cedar* (Vancouver: Douglas & McIntyre, 1984); Nancy Turner, *Plants in British Columbia Indian Technology* (Victoria, BC: Royal B.C. Museum, 1979). On camas and wapato (ska'us), see Terry Glavin, *The Last Great Sea: A Voyage through the Human and Natural History of the North Pacific Ocean* (Vancouver: Greystone/Douglas & McIntyre, 2000). For early botanical collecting in the Sitka Biome: Clive L. Justice, *Mr. Menzies' Garden Legacy: Plant Collecting on the Northwest Coast* (Vancouver: Cavendish Books, 2000). And on beetle infestations in forests: Timothy Egan, "As trees die, some cite the climate," *The New York Times,* June 25, 2002; Terry Glavin, "Towns on beetles' menu," *The Georgia Straight,* August 18, 2005.

On our "hard-wired" emotional affiliation with other living things, see Steven Pinker, *How the Mind Works* (New York: W.W. Norton, 1997); E.O. Wilson and Stephen R. Kellert, eds., *The Biophilia Hypothesis* (Washington, DC: Island Press, 1993).

For the alula, see K.R. Wood, "The Brighamia of Hawai'i," Conservation Department, National Tropical Botanical Garden, Lawa'i, Kaua'i, Hawai'i, available at: www.wildlifebiz.org/bellamy_good_news/34.asp. The story of Kew Gardens is told by David Blomfield, *The Story of Kew: The Gardens, The Village, the National Archives* (London: Leyborne Publications, 2003); Paul Cloutman, "Royal Botanical Gardens Kew: A Souvenir Guide" (London: The Trustees, Royal Botanical Gardens Kew, 2001); Pat Griggs, H.D.V. Prendergast, and Naomi Rumball, "Plants and People: An Exhibition of Items from the Economic Botany Collections" (London: The Trustees, Royal Botanical Gardens Kew, n.d.). On the coco-de-mer, see Fred Bruemmer, "Of Monstrous Moles and Unicorn Horns," *International Wildlife* 5, no. 15, 1998.

The Vavilov Institute is discussed in Mark MacKinnon, "Kremlin's decree sows seeds of discontent," *The Globe and Mail,* May 30, 2003; Galina Stolyarova, "Kremlin sets its sights on St. Isaac's Square," *Saint Petersburg Times,* January 28, 2003. On Vavilov himself, see "Nikolai Ivanovich Vavilov," *Seed News* 26, Summer 1999.

The scale of loss in diversity in food varieties is from the above-cited Mooney texts; for losses in domesticated-breed diversity, see Sustainable Development Department, Food and Agriculture Organization, 1998.

The potato blight and its history are described in G. Wilbert, et al., "The Population Structure of *Phytophthora infestans* from the Toluca Valley of Central Mexico Suggests Genetic Differentiation between Populations for Cultivated Potato and Wild *Solanum spp*," *Phytopathology*, April 2003. "Primitive" and wild cousins of food crops provide resistance to various moulds and blights, as told in Fowler and Mooney, 1990; Mooney, et al., 1988.

The Irish famine is eloquently described in Donald McKay, *Flight from Famine: The Coming of the Irish to Canada* (Toronto: McClelland & Stewart, 1992). Conditions in Ireland prior to the famine are discussed in George Cornwall Lewis, *Local Disturbances in Ireland* (Cork: Tower Books, 1977). On the contemporary importance of potato-crop diversity, see Blaine P. Friedlander, Jr., "CU and Polish scientists are leading effort to save valuable genetic archive," *Cornell Chronicle*, July 27, 2000.

For the Green Revolution, see Mooney's texts, along with Richard Manning, *Against the Grain: How Agriculture Has Hijacked Civilization* (New York: North Point Press/Farrar, Straus & Giroux, 2004); Richard Manning, "The Oil We Eat," *Harper's Magazine*, February 2004.

On biotechnology as a solution to the crisis in food-crop diversity, see George Monbiot, "Starved of the truth: Biotech firms are out to corner the market, so they have to persuade us something else is at stake," *The Guardian*, March 9, 2004. On declines in food quality and in standard of living, as well as responses to generic food worldwide, see "Filling the World's Belly," special supplement, *The Economist*, December 13, 2003; Eric Schlosser, *Fast Food Nation: The Dark Side of the All-American Meal* (New York: Houghton Mifflin, 2001). The ongoing saga of Percy Schmeiser can be followed at www.percyschmeiser.com. Also, see Damian Grammaticas, "Taiwan farmers in mass protest," BBC News, November 23, 2002; Robin McKie, "Fear of extremists kills off GM tests: Threat to dig up experimental crops drives British research overseas," *The Observer*, March 20, 2005; Reuters New Service, "Court backs U.K. activists in 'McLibel' case: European court overturns conviction for libeling U.S. fast-food chain," February 15, 2005; P. Sainath, "When Farmers Die," *India Together*, June 2004; Vandana Shiva, "The Suicide Economy of Corporate Globalisation," April 5, 2004, www.countercurrents.org/glo-shiva050404.htm; A.R. Vasvavi, "Loss of the Local and Spectres of the Global," *Leisa*, July 2001.

Alan Phillips's remarks are in Katy Guest, "Forbidden fruit: Red, round and tasteless? Not these beauties," *The Independent*, August 19, 2003.

The Singing Tree of Chungliyimti

On the Naga truce, see Luke Harding and Yoga Rangatia, "Naga rebels declare end of war with India," *The Guardian,* January 14, 2003. The overlapping maps that show Nagaland as a key centre for diversities in natural selection, artificial selection, language, and culture are in Tove Skutnabb-Kangas, et al., "Sharing a World of Difference: The Earth's Linguistic, Cultural and Biological Diversity" (Paris: UNESCO/World Wide Fund for Nature/Terralingua, 2003). The Conservation International "hotspot" maps, especially the rich biological diversity of the Himalayan region, are at www.biodiversityhotspots.org/xp/Hotspots/himalaya and in Eric Wikramanayake, et al., *Terrestrial Ecoregions of the Indo-Pacific: A Conservation Assessment* (Washington, DC: Island Press, 2000).

On Naga languages, see B.K. Boruah, *Nagamese: The Language of Nagaland* (New Delhi: Mittal Publications, 1993); Raymond G. Gordon, Jr., ed., *Ethnologue: Languages of the World* (Dallas, TX: SIL International, 2005); Rajesh Verma, "Northeastern India: A Lingustic Scenario" (New Delhi: Government of India Press Information Bureau, n.d.).

On the diversity of Nagaland vegetable varieties, see A.K. Mishra and U.C. Sharma, "Traditional Wisdom in Range Management for Resource and Environment Conservation in the North Eastern Region of India," *ENVIS Bulletin: Himalayan Ecology and Development* 9, no. 1 (2001); R.C. Upadhyay and R.C. Sundriyal, "Crop Gene Pools in the Northeast Indian Himalayas and Threats," in T. Partap and B. Sthapit, eds., *Managing Agrobiodiversity: Farmers' Changing Perspectives and Institutional Responses in the Hindu Kush-Himalayan Region* (Hertfordshire, UK: Earthprint Ltd. for the United Nations Environment Programme, 1998). For "150 species grown in the home gardens of just three villages" in Konyak Naga territory, see Earth Love Fund, 9 Market Place, Cirencester, Gloucestershire, GL7 2NX, U.K.

On Nagaland population growth, see Nirmalya Banerjee, "Record population growth in Nagaland," *The Times of India,* August 14, 2001.

See IUCN data for various species described. See also, on the golden langur, Jihosuo Biswas, "Evaluation of Population Status, Demography and Threats of Golden Langur, *Trachypithecus geei* (Khajuria, 1956) in different Fragmented Forests of Assam, India and Issues Related to Its Conservation" (Wisconsin Primate Research Center Library, 2003). On the sangai: Salam Rajesh, "A cry in the wilderness," *e-pao,* August 20, 2005, www.e-pao.net/index.html. On the leaf muntjac: A. Rabinowitz et al., "Description of the Leaf Deer, *Muntiacus putaoensis,* a New Species of Muntjac from Northern Myanmar," *Journal of Zoology* 249 (1999).

On the Naga insurgency, with specific reference to Khonoma, see Easterine Iralu, *A Naga Village Remembered* (Kohima, India: N.V. Press, 2003); Kaka D. Iralu, *Nagaland and India: The Blood and the Tears* (Kohima, India: N.V. Press, 2000); Pieter Steyn, *Zapuphizo: Voice of the Nagas* (Hong Kong: Centre of Asian Studies, University of Hong Kong, 1996).

For the *mithun:* S. Hedges, *Bos frontalis,* in IUCN, Red List of Threatened Species (2000), www.redlist.org.

For Nagaland's reforestation initiatives, see Curt Labond, "Promoting Sustainable Agroforestry in Nagaland," IDRC Reports (1999), web.idrc.ca/es/ev-5154-201-1-DO_TOPIC.html. See also a 1999 analysis of the outcome of the reforestation initiative undertaken by the Nagaland Environmental Protection and Economic Development project, at www.idrc.ca/uploads/user-S/10504294320Nagaland_Environmental_Protection_and_Economic_Development_Project_A_Self-Assessment_Using_Outcome_Mapping.htm. On tropical deforestation generally, see Daan van Soest, *Tropical Deforestation: An Economic Perspective* (Netherlands: Labyrinth Publications, 1998). And on *jhum* deforestation in Nagaland: Majid Husain, *Nagaland: Habitat, Society and Shifting Cultivation* (New Delhi: Rima Publishing House, 1988).

The Naga adoption of the Baptist faith is described in C. Walu Walling, *Sacrifice and Salvation in Ao-Naga Tradition: A Theological Perspective* (Impur, India: C. Walu Walling, 1997). For various Naga ethnographies and the Naga insurgency in relation to conflicts in India's northeast, see Sanjib Baruah, "Confronting Constructionism: Ending India's Naga War," *Journal of Peace Research* 40, no. 3 (May 2003); Partha S. Ghosh, "Ethnic and Religious Conflicts in South Asia," *Conflict Studies* 178 (1985); Syed Zarir Hussain, "India: The lost Jews of Mizoram," *Indo-Asian News Service,* December 16, 2003; J.H. Hutton, *The Angami Nagas* (London: Macmillan, 1921); International Working Group for Indigenous Affairs, "The Naga Nation and Its Struggle against Genocide," IWGIA Document 56 (Copenhagen: IWGIA, 1986); Kristoffel Leiten, "India: Multiple Conflicts in Northeast India," in Monique van Mekenkamp, Paul Tongeren, and Hans van de Veen, eds., *Searching for Peace in Central and South Asia* (Boulder, CO: Lynne Rienner, 2002); J.P. Mills, *The Ao Nagas* (London: Macmillan, 1926); Christoph Von

Furer-Haimendorf, *Naked Nagas* (Calcutta: Thacker Spink & Co., 1939).

On tigers in Burma (Myanmar), see Alan Rabinowitz, "Valley of Death," *National Geographic,* April 2004.

Epilogue: The Revenge of Kali

The epigraph is taken from E.O. Wilson, *Consilience: The Unity of Knowledge* (New York: Random House, 1999), 281.

A useful reference for Kali is Ajit Mookerjee, *Kali: The Feminine Force* (Rochester, VT: Destiny Books, 1988). On the rise of Shiv Sena, see Larissa McFarquhar, "The Strongman: Where Is Hindu-Nationalist Violence Heading?" *The New Yorker,* May 26, 2003.

Deaths every year due to extreme poverty: Jeffrey D. Sachs, *The End of Poverty: Economic Possibilities for Our Time* (New York: Penguin, 2005). One-third of humanity lives on less than $2 a day: Renato Ruggiero, Director-General, World Trade Organization, Opening remarks to the high level symposium on Trade and Development, March 17, 1999, WTO, Geneva, Switzerland. *The Times of India* editorial proposing anti-depressants for suicidal Indian farmers: S. Anklesaria Aiyar Swaminanthan, "Everybody loves farm suicides," *The Times of India,* August 1, 2004.

Parallels between twenty-first-century global civilization and the final days of the Sumerians, Mayans, and Romans are made by Ronald Wright, *A Short History of Progress* (Toronto: House of Anansi Press, 2004). "The choice between transcendentalism and empiricism ...": Wilson, *Consilience,* 262. The Montreal Protocol on Substances That Deplete the Ozone Layer: United Nations Development Programme, "The Vienna Convention and the Montreal Protocol," www.undp.org/seed/eap/montreal/montreal.htm. David Quammen's prospects for humanity's extinction, in "Planet of Weeds," *Harper's Magazine,* October 1998.

Index

M

N